PARADE OF LIFE
Monerans, Protists, Fungi, and Plants

Anthea Maton
Former NSTA National Coordinator
Project Scope, Sequence, Coordination
Washington, DC

Jean Hopkins
Science Instructor and Department Chairperson
John H. Wood Middle School
San Antonio, Texas

Susan Johnson
Professor of Biology
Ball State University
Muncie, Indiana

David LaHart
Senior Instructor
Florida Solar Energy Center
Cape Canaveral, Florida

Maryanna Quon Warner
Science Instructor
Del Dios Middle School
Escondido, California

Jill D. Wright
Professor of Science Education
Director of International Field Programs
University of Pittsburgh
Pittsburgh, Pennsylvania

Prentice Hall
Englewood Cliffs, New Jersey
Needham, Massachusetts

Prentice Hall Science

Parade of Life: Monerans, Protists, Fungi, and Plants

Student Text and Annotated Teacher's Edition
Laboratory Manual
Teacher's Resource Package
Teacher's Desk Reference
Computer Test Bank
Teaching Transparencies
Product Testing Activities
Computer Courseware
Video and Interactive Video

The illustration on the cover, rendered by Keith Kasnot, provides a glimpse of a few of the many organisms in a forest ecosystem.

Credits begin on page 193.

SECOND EDITION

© 1994, 1993 by Prentice-Hall, Inc., Englewood Cliffs, New Jersey 07632.

ISBN 0-13-225590-1

3 4 5 6 7 8 9 10   97 96 95 94

Prentice Hall
A Division of Simon & Schuster
Englewood Cliffs, New Jersey 07632

STAFF CREDITS

Editorial:	Harry Bakalian, Pamela E. Hirschfeld, Maureen Grassi, Robert P. Letendre, Elisa Mui Eiger, Lorraine Smith-Phelan, Christine A. Caputo
Design:	AnnMarie Roselli, Carmela Pereira, Susan Walrath, Leslie Osher, Art Soares
Production:	Suse F. Bell, Joan McCulley, Elizabeth Torjussen, Christina Burghard
Photo Research:	Libby Forsyth, Emily Rose, Martha Conway
Publishing Technology:	Andrew Grey Bommarito, Deborah Jones, Monduane Harris, Michael Colucci, Gregory Myers, Cleasta Wilburn
Marketing:	Andrew Socha, Victoria Willows
Pre-Press Production:	Laura Sanderson, Kathryn Dix, Denise Herckenrath
Manufacturing:	Rhett Conklin, Gertrude Szyferblatt

Consultants

Kathy French	National Science Consultant
Jeannie Dennard	National Science Consultant
Brenda Underwood	National Science Consultant
Janelle Conarton	National Science Consultant

CONTENTS

PARADE OF LIFE: MONERANS, PROTISTS, FUNGI, AND PLANTS

Activity Bank/Reference Section

Features

CONCEPT MAPPING

Throughout your study of science, you will learn a variety of terms, facts, figures, and concepts. Each new topic you encounter will provide its own collection of words and ideas—which, at times, you may think seem endless. But each of the ideas within a particular topic is related in some way to the others. No concept in science is isolated. Thus it will help you to understand the topic if you see the whole picture; that is, the interconnectedness of all the individual terms and ideas. This is a much more effective and satisfying way of learning than memorizing separate facts.

Actually, this should be a rather familiar process for you. Although you may not think about it in this way, you analyze many of the elements in your daily life by looking for relationships or connections. For example, when you look at a collection of flowers, you may divide them into groups: roses, carnations, and daisies. You may then associate colors with these flowers: red, pink, and white. The general topic is flowers. The subtopic is types of flowers. And the colors are specific terms that describe flowers. A topic makes more sense and is more easily understood if you understand how it is broken down into individual ideas and how these ideas are related to one another and to the entire topic.

It is often helpful to organize information visually so that you can see how it all fits together. One technique for describing related ideas is called a **concept map**. In a concept map, an idea is represented by a word or phrase enclosed in a box. There are several ideas in any concept map. A connection between two ideas is made with a line. A word or two that describes the connection is written on or near the line. The general topic is located at the top of the map. That topic is then broken down into subtopics, or more specific ideas, by branching lines. The most specific topics are located at the bottom of the map.

To construct a concept map, first identify the important ideas or key terms in the chapter or section. Do not try to include too much information. Use your judgment as to what is really important. Write the general topic at the top of your map. Let's use an example to help illustrate this process. Suppose you decide that the key terms in a section you are reading are School, Living Things, Language Arts, Subtraction, Grammar, Mathematics, Experiments, Papers, Science, Addition, Novels. The general topic is School. Write and enclose this word in a box at the top of your map.

SCHOOL

Now choose the subtopics—Language Arts, Science, Mathematics. Figure out how they are related to the topic. Add these words to your map. Continue this procedure until you have included all the important ideas and terms. Then use lines to make the appropriate connections between ideas and terms. Don't forget to write a word or two on or near the connecting line to describe the nature of the connection.

Do not be concerned if you have to redraw your map (perhaps several times!) before you show all the important connections clearly. If, for example, you write papers for Science as well as for Language Arts, you may want to place these two subjects next to each other so that the lines do not overlap.

One more thing you should know about concept mapping: Concepts can be correctly mapped in many different ways. In fact, it is unlikely that any two people will draw identical concept maps for a complex topic. Thus there is no one correct concept map for any topic! Even

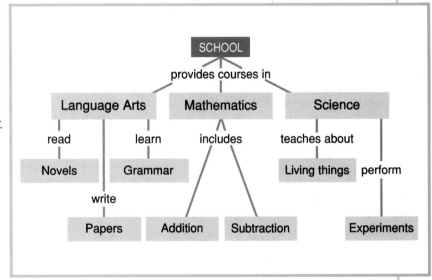

though your concept map may not match those of your classmates, it will be correct as long as it shows the most important concepts and the clear relationships among them. Your concept map will also be correct if it has meaning to you and if it helps you understand the material you are reading. A concept map should be so clear that if some of the terms are erased, the missing terms could easily be filled in by following the logic of the concept map.

PARADE OF LIFE

Monerans, Protists, Fungi, and Plants

▼ *The first marchers to appear in the parade of life were monerans, more commonly known as bacteria.*

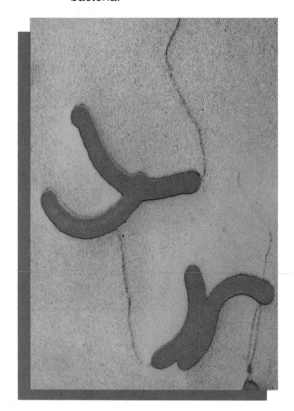

Think about the last time you saw a parade. You may have seen marching bands, beautiful floats, enormous balloons, showers of ticker tape, or waving flags.

At this moment, you are in the middle of the biggest, oldest, most spectacular parade on Earth—the parade of life. This parade began its journey through time more than 3.5 billion years ago, and its marchers are all the living things on Earth—past, present, and future. In this book, you will begin your exploration of the parade of life—an adventure that will take you all over the world.

Two billion years after bacteria started the parade of life, the next group of marchers appeared. This group consists of living things known as protists. Protists, such as wheel-shaped diatoms, are made of a single, complex cell. ▼

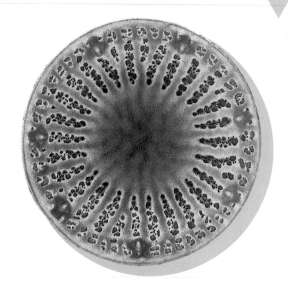

CHAPTERS

Flowering plants are newcomers to the parade of life. Many flowers are actually clusters of flowers. Each yellow bump in the center and each "petal" of a painted daisy is a single tiny flower.

Living things that are composed of many cells are relatively recent additions to the parade of life. Many-celled living things include fungi, such as mushrooms, as well as plants.

Discovery *Activity*

Taking a Closer Look

1. Obtain a sample of pond water. What does the pond water look like? Can you see living things in the water?

2. Using a medicine dropper, place a drop of the pond water in the center of a glass microscope slide. Cover the drop with a coverslip.

3. Examine the drop of water with a microscope under low and high powers. What do you observe?

   ■ How do you think the inventor of the microscope felt when he first looked at pond water with his invention?

   ■ What do your observations tell you about some of the living things that march alongside you in the parade of life?

   ■ Why are microscopes necessary tools for studying the parade of life?

Classification of Living Things

Guide for Reading

After you read the following sections, you will be able to

1–1 History of Classification

- Give examples of the ways classification is used in science and in everyday life.
- Explain how binomial nomenclature is used to name living things.

1–2 Classification Today

- Relate biological classification to evolution.
- List the seven major classification groups.

1–3 The Five Kingdoms

- Describe some general characteristics of each of the five kingdoms.

Mary Courtnay-Latimer had never seen anything like the creature the crew of a fishing boat had captured near the mouth of the Chalumna River in South Africa. The monstrous fish stretched more than 1.5 meters, or about six times the length of this page. Large spiny steel-blue scales covered its oily body. A powerful jaw hung down from a frightening face. Most peculiar of all, its fins were attached to what appeared to be stubby legs!

After searching through many books but finding no description of the strange fish, Latimer remembered someone she thought might be able to solve the riddle. That someone was Dr. James L. B. Smith, a well-known fish expert at a nearby university.

The scientist was shocked. He later wrote, "I would hardly have been more surprised if I met a dinosaur on the street." Smith was looking at an animal thought to have become extinct more than 60 million years ago. Yet, in a flash, he had been able to identify the fish as a coelacanth (SEE-luh-kanth). Smith gave the coelacanth its scientific name: *Latimeria chalumnae.* To do so, he used biological classification—a special system that helps scientists identify and name organisms. And it is this special system that you will learn about in the pages that follow.

Journal *Activity*

You and Your World You probably know someone who often puts people into categories such as "jocks" or "brains." In your journal, discuss the practice of putting people into categories.

◄ *The coelacanth is the only surviving member of the ancient group of fishes from which modern four-footed land animals are thought to have evolved.*

1-1 History of Classification

Thousands of years ago, as people made observations about the world around them, they began to recognize that there were different groups of living things. There were animals and there were plants. Of the animals, some had claws and sharp teeth and roamed the land. Others had feathers and beaks and flew in the air. Still others had scales and fins and swam in the water.

Plants too showed a wide range of differences. Not only did they vary in shape, size, and color, but some were good to eat whereas others were poisonous. People soon learned that poisonous plants were best avoided. In a similar way, people learned that some animals, such as those with sharp teeth, were dangerous. Others, such as those with feathers, were relatively harmless.

What these early people were doing is something you often do in your daily life. They were giving order to the world around them by putting things in groups or categories based on certain characteristics. In other words, they were developing simple systems of classification. **Classification is the grouping of things according to similar characteristics.**

Stop and think for a moment about the ways in which you classify things every day. Perhaps you group your clothing by season—lightweight, cool items for summer and heavy, warm items for winter.

Figure 1–1 *When humans first came to North America tens of thousands of years ago, they found a number of unusual animals. How did classifying these animals help humans survive?*

Or perhaps by type—pants, sweaters, skirts, jackets. What are some other things that you sort in a meaningful way?

Classification is important to all fields of science, not just to the subject of biology you are now studying. For example, geologists classify rocks, soils, and fossils. Meteorologists (people who study the weather) classify clouds, winds, and types of storms. And all of the 109 known chemical elements are classified into a system that helps chemists understand their behavior.

Classification is important in subjects other than science. In English, parts of speech are categorized as nouns, verbs, adjectives, and adverbs, to name a few. In mathematics, you work with odd numbers and even numbers, circles, rectangles, and triangles. In history, you group people and events according to time periods or geographic locations. You know that music can be rock-and-roll, rhythm and blues, country and western, or classical.

These are but a few examples of the important role that classification plays in all phases of life. For the people living thousands of years ago, classifying living things according to observable characteristics often helped them to survive. For you, classifying objects probably makes life easier and more meaningful. For scientists, classification systems provide a

ACTIVITY
DOING

Classification of Rocks

1. Obtain ten different rocks.

2. Examine the rocks. Notice how they are similar and how they are different.

3. Decide which characteristics of the rocks are most important. Use these characteristics to create a classification system for the rocks.

4. Notice that in this case there is no single correct way to classify your rocks. However, geologists do classify rocks in a particular way. Some characteristics that are important to geologists are the ways rocks are formed, the kind of chemicals that make up the rocks, and the shape of the crystals in the rocks.

Figure 1–2 *Geologists classify minerals according to the chemical compounds that make them up. Malachite (top) and azurite (bottom) are both forms of copper carbonate.*

means of learning more about life on Earth and of discovering the special relationships that exist between different kinds of living things.

But no matter who is doing the classifying or what is being classified, a classification system is always based on observable characteristics a group of things share. Good classification systems are meaningful, easily understood, and readily communicated among people.

Biological classification systems name and organize living things in a logical, meaningful way. To date, scientists have identified more than 2.5 million different types of living things—and their job is not even close to being finished! Some biologists estimate that there may be at least another 7 million different kinds of organisms living in tropical rain forests and in the depths of the Earth's oceans. In order to bring some order to this great diversity of living things, biologists have developed systems of classification.

The science of classification is a branch of biology known as taxonomy (taks-AH-nuh-mee). Scientists

Figure 1–3 *The dizzying variety of corals, fishes, and algae in a coral reef represents only a tiny fraction of the Earth's living things. Why is it necessary for biologists to classify living things?*

Figure 1–4 *The instruments in a marching band can be classified as woodwinds, percussion, and brass. How are some other everyday things classified?*

who work in this field are called taxonomists. Taxonomy has a long history, during which many classification systems were developed, changed to fit new facts and theories, and even rejected and replaced with better systems. This fact should not surprise you if you remember that science is an ongoing process that is often marked by change. This ability to change when new knowledge becomes available is one of science's greatest strengths.

The First Classification Systems

In the fourth century BC, the Greek philosopher Aristotle proposed a system to classify living things. He divided organisms into two groups: plants and animals. He also placed animals into three groups according to the way they moved. One group included all animals that flew, another group included all those that swam, and a third group included all those that walked.

Although this system was useful, it had some problems. Can you see why Aristotle's system for classifying animals is not perfect? It ignores the ways in which animals are similar and different in form. According to Aristotle, both a bird and a bat would be placed in the same group of flying animals. Yet in some basic ways, birds and bats are quite different. For example, birds are covered with feathers, whereas bats are covered with hair.

Figure 1–5 *Although the false vampire bat can fly like a bird, it belongs to the same class as the mouse it is about to eat (left). The egret belongs to a separate class of animals (right). How are the wings and other characteristics of the bat and egret different?*

Although the system devised by Aristotle would not satisfy today's taxonomists, it was one of the first attempts to develop a scientific and orderly system of classification. In fact, Aristotle's classification system was used for almost 2000 years. And until the middle of the twentieth century, scientists continued to use Aristotle's system of classifying living things as either plants or animals.

By the seventeenth century, biologists had started to classify organisms according to similarities in form and structure. They examined the organism's internal anatomy as well as its outward appearance. This helped them to place animals and other organisms into groups that were more meaningful than those Aristotle had created.

The classification system we use today is based on the work of the eighteenth-century Swedish scientist Carolus Linnaeus. Linnaeus built upon the work of previous scientists to develop his new system of classification. Like Aristotle, Linnaeus identified all living things as either plants or animals. Like the seventeenth-century biologists, he grouped plants and animals according to similarities in form. And like almost all other previous taxonomists, he used a system that consisted of groups within larger groups within still larger groups. Linnaeus spent the major part of his life using his classification system to describe all known plants and animals.

Linnaeus also developed a simple system for naming organisms. His system is such a logical and easy way of naming organisms that it is still used today.

Naming Living Things

Before Linnaeus developed his naming system, plants and animals had been identified by a series of Latin words. These words, sometimes numbering as many as 12 for one organism, described the physical features or appearance of the organism. To make things even more confusing, the names of plants and animals were rarely the same from book to book or from place to place.

BINOMIAL NOMENCLATURE   The naming system devised by Linnaeus is called **binomial nomenclature** (bigh-NOH-mee-uhl NOH-muhn-klay-cher). In this system, each organism is given two names—which is exactly what the term binomial (consisting of two names) nomenclature (system of naming) means. The two names are a **genus** (plural: genera) name and a **species** name.

To help you understand this method of naming, think of the genus name as your family name. The species name could then be thought of as your first name. Your family name and your first name are the two names that identify you—the two names that most people know you by. They represent the most specific way of identifying you by name.

NAMING ORGANISMS TODAY   Each kind of organism is given its very own two-part scientific name. An organism's scientific name is made up of its genus name and species name. The genus name is capitalized, but the species name begins with a small letter. Both names are printed in italics, which will help you recognize scientific names as you do your reading. Here is an example: The genus and species name for a wolf is *Canis lupus*. These two names identify this organism—which, by the way, has become very rare in the wild. Although most scientific names are in Latin, some are in Greek. Now think about the scientific name of the honeybee, *Apis mellifera*. What is the honeybee's genus name? Its species name?

Each organism has only one scientific name. And no two organisms can have the same scientific name. To understand why this is important, consider the following story. In North and South America, a certain large cat is called a mountain lion by some people, a cougar by others, and a puma by still others.

Figure 1–6 *The honeybee was once called* Apis pubescens, thorace subgriseo, abdomine fusco, pedibus posticis glabris utrinque margine ciliatus. *This means "fuzzy bee, light gray middle, brown body, smooth hind legs that have a small bag edged with tiny hairs." Linnaeus named the honeybee* Apis mellifera, *which means "honey-bearing bee."*

ACTIVITY DOING

All in the Family

Use reference books to learn about the various families in the plant and animal kingdoms. Choose the family that you find most interesting. Make a collage of the family using pictures from old magazines and newspapers. Present the collage to your class.

If these people were to talk to one another about this animal, they might get rather confused, thinking they were talking about three different animals. But scientists cannot afford to have such confusion. So scientists throughout the world know this large cat by only one name, *Felis concolor*. This name easily identifies the cat to all scientists, no matter where they live or what language they speak.

Keep in mind that it is not necessary to memorize the scientific names of different organisms. Even biologists know only a few names by heart, and most of these names are of organisms they have studied for years. What is important for you to know is that each organism has a scientific name that is used and understood all over the world, and that this name is related to the way the organism is classified.

Figure 1–7 *The wolf and the puma are known by many different common names. However, each organism has its own unique scientific name. What is the scientific name for the wolf? The puma?*

1–1 Section Review

1. Describe some ways in which you use classification in everyday life.
2. What is taxonomy?
3. What is binomial nomenclature? How is it used?

Critical Thinking—*Evaluating Systems*
4. Discuss three problems with the classification and naming systems that existed before Linnaeus.

1–2 Classification Today

In the 200 years since Linnaeus developed his classification and naming systems, knowledge of the living world has grown enormously. And as the understanding of organisms improved, it became necessary to adjust the system of biological classification. Two things in particular have had a large effect on biological classification. One of these is Charles Darwin's theory of evolution. The other is advances in technology that have enabled scientists to take a better look at organisms.

Evolution and Classification

As you can see in Figure 1–8, wolves and lions both developed from a meat-eating animal that existed about 60 million years ago. And lions and house cats both developed from a catlike animal that lived about 15 million years ago. During the long history of life on Earth, organisms have changed, or evolved. You can think of evolution as the process in

Guide for Reading

Focus on this question as you read.

▶ *What are the classification groups from largest to smallest?*

Figure 1–8 *Because they evolved from a shared ancestor, lions, cats, wolves, and the catlike animal belong to the same classification group (order Carnivora). Cats and lions also belong to a smaller classification group (family Felidae), which contains all the descendants of the catlike animal. Why aren't wolves placed in the family Felidae?*

Meat-eating animal

Catlike animal

House cat

Wolf

Lion

Time (millions of years ago)

60

45

0

Figure 1–9 *As knowledge of the evolutionary relationships among animals improved, the lesser panda (bottom) and the giant panda (top center) were classified and reclassified. The data now available support a classification scheme in which the giant panda belongs to the same family as other bears, such as the grizzly (top left). The lesser panda is placed in the same family as raccoons (top right).*

which new kinds of organisms develop from previously existing kinds of organisms.

Evolutionary relationships, such as those between the ancient catlike animal and house cats or between wolves and lions, are extremely important to modern taxonomists. Evolutionary relationships are the basis for the modern system of biological classification. Modern taxonomists try to classify living things in such a way that each classification group contains organisms that evolved from the same ancestor.

The knowledge of evolution has changed the nature of biological classification groups. It has also changed the job of taxonomists. Because they did not know about evolution, taxonomists of the past felt free to classify organisms in any manner that made sense to them. They did not classify prehistoric organisms because they did not know about them. And they could choose any characteristics that they thought were important as their basis for classification.

Present-day taxonomists, on the other hand, classify organisms in a way that shows evolutionary relationships. They must consider organisms that existed in the distant past as well as those that exist in the present. And they classify organisms using characteristics that are proven to be good indicators of evolutionary relationships.

Technology and Classification

Today, 200 years after Linnaeus completed his work, scientists consider many factors when classifying organisms. Of course, they still examine the large internal and external structures, but they also rely on other observations. The invention of the microscope has allowed scientists to examine tiny structures hidden within the cells of an organism. It has also allowed them to examine organisms at their earliest stages of development. And special chemical tests have been developed that enable scientists to analyze the chemical building blocks of all living things. All of these techniques are important tools that help scientists group and name organisms.

Classification Groups

At first glance, the modern classification system may seem complicated to you. However, it is really quite simple, especially when you keep in mind its purpose. The system of classification used today does two jobs. First, it gives each organism a unique name that scientists all over the world can use and understand. Second, it groups organisms according to basic characteristics that reflect their evolutionary relationships.

All living things are classified into seven major groups: kingdom, phylum, class, order, family, genus, and species. The largest and most general group is the kingdom. For example, all animals belong to the animal kingdom. The second largest group is the phylum (FIGH-luhm; plural: phyla). A phylum includes a large number of very different organisms. However, these organisms share some important characteristics. A species is the smallest and most specific group in the classification system. Members of the same species share many characteristics and

ACTIVITY
WRITING

A Secret Code

You can remember the correct classification sequence from the largest to the smallest group by remembering this sentence: Kings play cards on fat green stools. The first letter of each word is the same as the first letter of each classification group.

Write two additional sentences that can help you to remember the classification sequence.

Figure 1–10 *The male (right) and female (left) grand eclectus parrots look so different that they were once thought to be separate species.*

CLASSIFICATION OF THE LION

Kingdom
Animalia

Phylum
Chordata

Class
Mammalia

Order
Carnivora

Family
Felidae

Genus
Panthera

Species
leo

Figure 1–11 *This chart shows several other organisms that are in the same classification groups as the lion. To what class do lions belong?*

are similar to one another in appearance and behavior. In addition, members of the same species can interbreed and produce offspring. These offspring can in turn produce offspring of their own.

Ideally, the largest classification groups represent the earliest ancestors and the most ancient branches of life's family tree. And the smallest classification groups contain organisms that evolved from shared ancestors that lived in the relatively recent past. But because people do not know everything there is to know about evolution, these categories are not perfect in real life. As scientists learn more and more about evolutionary relationships and about the history of life on Earth, the classification system is changed. Sometimes the changes are tiny. Other times, the changes are quite large. And once in a while, taxonomists choose to keep an old group because it is particularly useful and logical, even if it does not perfectly reflect evolutionary history.

Biologists often think of these classification groups as forming a tree in which the trunk represents the kingdom, the main branches represent the phyla within the kingdom, and the tiny twigs at the tips of the branches represent species. You can also

Figure 1–12 *The relationships among classification groups can be represented as a tree. This classification tree shows the major groups of animals. What animals are in the same phylum as the centipede and the red beetle?*

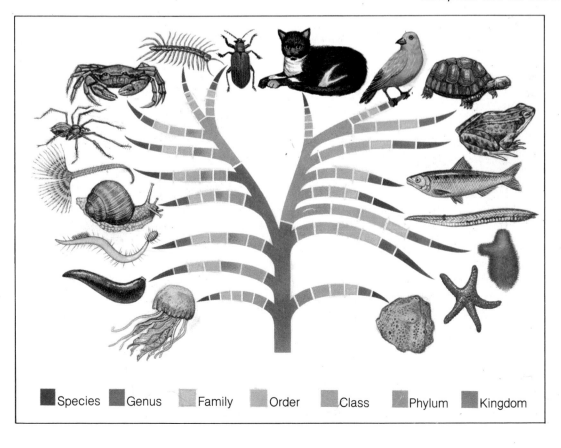

Species Genus Family Order Class Phylum Kingdom

PROBLEM
Solving

Classifying the Dragons of Planet Nitram

Imagine that you are a famous space explorer and biologist. You have recently arrived on the planet Nitram. Your job is to study the Nitramian animals and develop a system for classifying them.

You decide to begin your work by classifying Nitram's "dragons." You think that this name is not very scientific. However, you must admit that the animals look a lot like the monsters of Earth legends.

You ask the computer to give you a brief summary of all the information it has on Nitramian dragons. The computer produces the following printout.

THE DRAGONS OF NITRAM

Drako: About 5 to 6 meters tall. Four legs. Two batlike wings. Able to fly. Body covered in scales. Lives in mountains. Feeds on large animals.

Quetzalcoatl: About 2 meters long. Two legs. Two birdlike wings. Able to fly. Green and red feathers. Long feathers of many colors on top of head and around neck. Green scales on snakelike tail. Lives in tropical jungles. Eats fruit and small animals.

Sanjorge: About 2 to 5 meters tall. Four legs. Two small batlike wings. Not able to fly. Body has reddish-brown and greenish-brown scales. Lives in mountains. Probably eats animals.

Smaug: Largest known dragon. About 6 to 9 meters tall. Four legs. Two batlike wings. Able to fly. Body red, belly orange. Seems to be covered in scales. Extremely aggressive. Lives in mountains. Feeds on large animals, including humans.

Tailoong: About 4 meters long. Long, thin, snakelike body covered with gold scales. Four legs. Lionlike mane of long, colorful feathers. Lives in forests near lakes. Feeds on flowers and water plants.

Wyvern: About 2 to 3 meters long. Two legs. Two birdlike wings. Able to fly. Head, neck, and tail covered with red, yellow, and brown scales. Long red feathers around base of neck. Wings and body covered with brown feathers. Lives in mountains. Feeds on small animals. May also feed on the remains of dead animals.

Classifying Animals

1. Develop a classification system for the Nitramian dragons. Explain how you devised your classification scheme.

2. Like the animals of Earth, dragons are the result of millions of years of evolution. Which types of dragons seem to be most closely related? Explain.

3. When a sanjorge gets to be about 5 meters tall, its wings begin to grow rapidly. Within a year or two, it can fly. At the same time, its color changes to either red or green. How do these new data affect your classification system?

think of the classification groups as boxes within boxes. For example, you might open a huge kingdom box to discover several large phyla boxes, each of which contains a number of still smaller class boxes. Each time you opened a box, you would find one or more smaller boxes. If you opened the smallest boxes (the species boxes), you would find a number of individuals all of the same type.

By knowing which branches to climb, you can eventually arrive at one particular twig. (Provided, of course, that the branches are strong enough to bear your weight!) If you know which boxes to open in a set of boxes within boxes, you will sooner or later find the one tiny box you are looking for. Similarly, if you go through each level of classification groups, you will finally arrive at one species; that is, one specific kind of organism.

1-2 Section Review

1. List the classification groups from largest to smallest.
2. What is evolution? How does evolution affect the way organisms are classified?

Critical Thinking—*Making Inferences*
3. Explain why knowing the classification of an unfamiliar organism can tell you a lot about that organism.

1-3 The Five Kingdoms

The discoveries of new living things and the changing ideas about the most effective ways of classifying life forms have resulted in the five-kingdom classification system we use now. **Today, the most generally accepted classification system contains five kingdoms: monerans, protists, fungi, plants, and animals.**

As is often the case in science, not all scientists agree on this classification system. And this is an important idea for you to keep in mind. More research may someday show that different systems make more

Guide for Reading

Focus on this question as you read.

▶ *What are the five kingdoms of living things?*

Activity Bank

A Key to the Puzzle, p.170

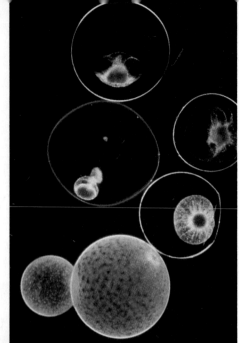

Figure 1–13 *Almost all members of the kingdoms Monera and Protista are microscopic. The blue-green bacterium* Anabaena, *a moneran, forms squiggly chains of cells. The glow-in-the-dark spheres and foamy starburst are types of protists. What is the main difference between monerans and protists?*

sense and better represent how living things evolved. But for now, this five-kingdom system is a useful tool for studying living things—and that's exactly what taxonomy is all about.

MONERANS Bacteria are placed in the kingdom Monera. Monerans are unicellular organisms, or organisms that consist of only one cell. A moneran's cell does not have its hereditary material enclosed in a nucleus, the structure that in other cells houses this important material. In addition to a nucleus, moneran cells lack many other structures found in other cells. Because of their unique characteristics, monerans are considered to be very distantly related to the other kingdoms.

Like other organisms, monerans can be placed into two categories based on how they obtain energy. Organisms that obtain energy by making their own food are called **autotrophs.** Such a name makes a lot of sense since the prefix *auto-* means self and the root word *-troph* means food. Organisms that cannot make their own food are called **heterotrophs.** The prefix *hetero-* means other. Can you explain why heterotroph is a good name for such organisms? Heterotrophs may eat autotrophs in order to obtain food or they may eat other heterotrophs. But all heterotrophs ultimately rely on autotrophs for food.

Scientists have evidence that monerans were the earliest life forms on Earth. They first appeared about 3.5 billion years ago.

PROTISTS The kingdom Protista includes most of the unicellular (one-celled) organisms that have a nucleus. The nucleus controls the functions of the cell and also contains the cell's hereditary material. In addition, the cell of a protist has special structures that perform specific functions for the cell.

A number of protists are capable of animallike movement but also have some distinctly plantlike characteristics. Specifically, they are green in color and can use the energy of light to make their own food from simple substances. As you can imagine, such organisms are difficult, if not impossible, to classify using a two-kingdom classification system. They are neither plants nor animals. Or perhaps they are both plants and animals! These puzzling organisms are one of the reasons scientists finally decided to abandon the two-kingdom system.

Protists were the first kind of cells that contain a nucleus. Ancient types of protists that lived millions and millions of years ago are probably the ancestors of fungi, plants, animals, and modern protists.

FUNGI As you might expect, the world's wide variety of fungi make up the kingdom Fungi. Most fungi are multicellular organisms, or organisms that consist of many cells. Although you may not realize it, you may be quite familiar with fungi. Mushrooms are fungi. So are the molds that sometimes grow on leftover foods that have remained too long in the refrigerator. And the mildews that may appear as small black spots in damp basements and bathrooms are also fungi.

Figure 1–14 *The mushroom (left), shelf fungi (top right), and starfish stinkhorn (bottom right) all belong to the kingdom Fungi. The brownish slime on which the flies are feasting contains the stinkhorn's spores. If the spores end up in a favorable place after being deposited in the flies' droppings, they will grow into new stinkhorns.*

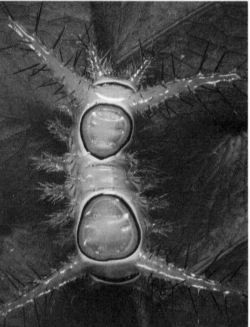

Figure 1–15 *The haired saddleback caterpillar and the pink katydid are both animals that feed on plants. How are animals similar to plants? How are they different?*

For many years, fungi were classified as plants. However, they are quite different from plants in some basic ways. Their cells fit together in a different way. Their cell wall, a tough protective layer that surrounds the cell, is made of a different substance than the cell wall of plants. And most importantly, plants are able to use the energy of light to make their own food from simple substances. Fungi cannot. Like animals, fungi must obtain their food and energy from another source. Are fungi autotrophs or heterotrophs?

PLANTS Plants make up the kingdom Plantae. Most members of the plant kingdom are multicellular (many-celled) autotrophs. You are probably quite familiar with members of this kingdom as it includes all the plants you have come to know by now— flowering plants, mosses, ferns, and certain algae, to name a few.

ANIMALS Animals are multicellular organisms that comprise the kingdom Animalia. Like other multicellular organisms such as plants and certain fungi, animals have specialized tissues, and most have organs and organ systems. Unlike plants, animals are heterotrophs.

1–3 Section Review

1. Name the five kingdoms of living things.
2. List three important characteristics for each of the five kingdoms.
3. How is the way an autotroph gets food different from the way a heterotroph gets food?

Connection—*Classifying Organisms*
4. Suppose that creatures from a distant planet are multicellular heterotrophs whose cells lack cell walls. Which kingdom of the Earth's organisms do these creatures most closely resemble?

CONNECTIONS

What's in a Name?

You may be familiar with the German *fairy tale* "Rumpelstiltskin." In this story, a magical dwarf saves a queen's life by helping her spin straw into gold. To pay the dwarf for his help, the queen must give him her baby—or guess the dwarf's secret name. In the end, she learns the dwarf's name. And everyone (except the dwarf) lives happily ever after.

Tales like this one reflect the ancient idea that names are extremely important. People once believed that knowing something's true name gave you magical power over it.

Knowing the true, or scientific, names of organisms is not going to give you magical power over them. But knowing what the scientific names mean can help give you a different kind of power—the power to figure out unfamiliar words.

As you look over the following examples of scientific names, remember that it is not important to memorize names. And don't let the strangeness of the names scare you. In time, you will become more familiar with scientific names and learn to see—and use—the patterns in them. You will discover that knowing the secrets of scientific names can be informative and even fun.

▲ *Naja melanoleuca* is a black-and-white-lipped cobra, a venomous snake. In India and Southeast Asia, a *naja,* or *naga,* is an imaginary beast that usually takes the form of a giant snake with a human head. A *naga* can also take the form of a human or a snake. (*melano-* means black; *leuca-* means white.)

▲ *Narcissus poeticus* is a sweet-smelling flower. According to a Greek myth, Narcissus was a handsome young man who fell in love with his own reflection. Eventually, the gods transformed Narcissus into a flower. In the spring, narcissus flowers may sometimes be found near a pond or stream. The long stems of the flowers may cause them to lean over the water so that it looks like they are admiring their own reflection. The story of Narcissus has added words other than the name of a flower to the English language. A person who is completely in love with himself (or herself) is called a narcissist. Linnaeus's name for the flower means "narcissus of the poets."

Laboratory Investigation

Whose Shoe Is That?

Problem
How can a group of objects be classified?

Materials *(per group of six)*

students' shoes
pencil and paper

Procedure

1. At your teacher's direction, remove your right shoe and place it on a work table.
2. As a group, think of a characteristic that will divide all six shoes into two kingdoms. For example, you may first divide the shoes by the characteristic of color into the brown shoe kingdom and the non-brown shoe kingdom.
3. Place the shoes into two separate piles based on the characteristic your group has selected.
4. Working only with those shoes in one kingdom, divide that kingdom into two groups based on a new characteristic. The brown shoe kingdom, for example, may be divided into shoes with rubber soles and shoes without rubber soles.
5. Further divide these groups into subgroups. For example, the shoes in the rubber-soled group may be separated into a shoelace group and a nonshoelace group.

6. Continue to divide the shoes by choosing new characteristics until you have only one shoe left in each group. Identify the person who owns this shoe.
7. Repeat this process working with the non-brown shoes.
8. Draw a diagram similar to the one shown to represent your classification system.

Observations

1. How many groups are there in your classification system?
2. Was there more than one way to divide the shoes into groups? How did you decide which classification groups to use?

Analysis and Conclusions

1. Was your shoe classification system accurate? Why or why not?
2. If brown and nonbrown shoe groups represent kingdoms, what do each of the other groups in your diagram represent?
3. Compare your classification system to the classification system used by most scientists today.
4. **On Your Own** Follow a similar procedure to classify all the objects in a closet or a drawer in your home.

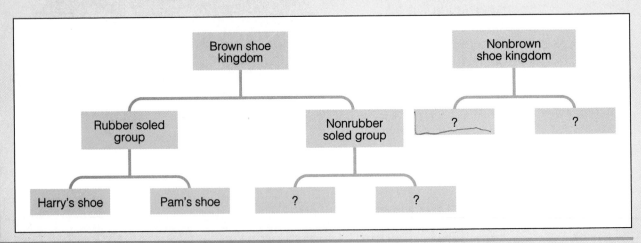

Summarizing Key Concepts

1–1 History of Classification

▲ Classification is the grouping of things according to similar characteristics.

▲Classification is important in all fields of science, in subjects other than science, and in everyday life.

▲ Good classification systems are meaningful, easily understood, and readily communicated among people.

▲ Classification systems organize and name living things in a logical, meaningful way.

▲ The science of biological classification is called taxonomy.

▲ Aristotle invented one of the first systems of biological classification. This system was used for about 2000 years.

▲ The classification and naming systems used today are based on the work of Carolus Linnaeus.

▲ In binomial nomenclature, which was invented by Linnaeus, each kind of organism is given a unique two-part name. This scientific name is used and understood all over the world. The scientific name is also related to the way the organism is classified.

1–2 Classification Today

▲ As our understanding of living things improves, it becomes necessary to revise our system of biological classification.

▲ Evolutionary relationships are the basis for the modern system of biological classification.

▲ Advances in technology have increased our knowledge of living things and thus have had an effect on how organisms are classified.

▲ The classification groups from largest to smallest are: kingdom, phylum, class, order, family, genus, and species. If you go through each level of classification groups, you finally arrive at one specific species.

▲ Each of an organism's classification groups tells you something about its characteristics.

1–3 The Five Kingdoms

▲ Today, the most generally accepted classification system contains five kingdoms: monerans, protists, fungi, plants, and animals.

▲ Autotrophs can make food from simple raw materials. Heterotrophs cannot make their own food.

Reviewing Key Terms

Define each term in a complete sentence.

1–1 History of Classification
binomial nomenclature
genus
species

1–3 The Five Kingdoms
autotroph
heterotroph

Chapter Review

Content Review

Multiple Choice

Choose the letter of the answer that best completes each statement.

1. Which of the following is not a characteristic of a good classification system?
   a. It shows relationships among objects.
   b. It is meaningful.
   c. It is readily communicated among people.
   d. It creates confusion.
2. The branch of biology that deals with naming and classifying organisms is called
   a. exobiology.
   b. taxonomy.
   c. phylum.
   d. binomial nomenclature.
3. The largest classification group is the
   a. species.
   b. order.
   c. phylum.
   d. kingdom.
4. A genus can be divided into
   a. phyla.
   b. orders.
   c. species.
   d. families.
5. Carolus Linnaeus classified plants and animals according to similarities in
   a. color.
   b. habits.
   c. structure.
   d. size.
6. Which organisms have cells that do not contain a nucleus?
   a. monerans
   b. fungi
   c. plants
   d. protists
7. Which of the following statements about animals in the same species is false?
   a. They evolved from a shared ancestor.
   b. They share certain characteristics.
   c. They can interbreed and produce offspring.
   d. They have identical characteristics.
8. Which organism is an autotroph?
   a. frog
   b. mushroom
   c. lion
   d. maple tree

True or False

If the statement is true, write "true." If it is false, change the underlined word or words to make the statement true.

1. In Linnaeus's classification system, the smallest group was the genus.
2. The classification groups from largest to smallest are: kingdom, phylum, class, family, order, species, genus.
3. Multicellular organisms are composed of only one cell.
4. In a five-kingdom classification system, bacteria are classified as plants.
5. Green plants are heterotrophs.
6. Autotrophs are organisms that cannot make their own food.
7. The science of classification is called taxonomy.
8. The first word in a scientific name is the species.

Concept Mapping

Complete the following concept map for Section 1–1. Refer to pages B6–B7 to construct a concept map for the entire chapter.

Concept Mastery

Discuss each of the following in a brief paragraph.

1. Why do scientists classify organisms?
2. What type of characteristics did Linnaeus use to develop his classification system of living things?
3. How is the classification system used by scientists today different from the classification system developed by Linnaeus? How is it similar?
4. Describe each of the kingdoms used in the five-kingdom classification system.

Give an example of an organism in each kingdom.
5. How do an autotroph and a heterotroph differ? Give an example of each.
6. How did the knowledge of evolution affect the way organisms are classified?
7. What are the two major jobs of the modern classification system? Explain why these jobs are important.

Critical Thinking and Problem Solving

Use the skills you have developed in this chapter to answer each of the following.

1. **Identifying relationships** Which two of the following three unicellular organisms are most closely related: *Entamoeba histolytica, Escherichia coli, Entamoeba coli?* Explain your answer.
2. **Developing a model** Design a classification system for objects that might be found in your closet. Then draw a diagram that illustrates your classification system.
3. **Making comparisons** In what ways were the classification and naming systems developed by Linnaeus an improvement over previous systems?
4. **Applying concepts** Why is it that scientists do not classify animals by what they eat or where they live?
5. **Relating cause and effect** Explain why advances in technology may change the way organisms are classified.
6. **Applying concepts** Suppose you discovered a new single-celled organism. This organism has a nucleus and a long taillike structure that it uses to move itself through the water in which it lives. It also has a large cup-shaped structure that is filled with a green substance. This green substance is involved in making food from simple substances. In what kingdom

would you place this organism? What are your reasons?
7. **Using the writing process** Some experts estimate that there are more unknown organisms in the tropical rain forests than there are known organisms in the world. These rain forests may be destroyed before the organisms in them can be studied and classified. It is possible that some of these organisms may be helpful in medicine, farming, and industry. Write a script for a television news program protesting the destruction of rain forests. Offer reasons why rain forests should be protected.

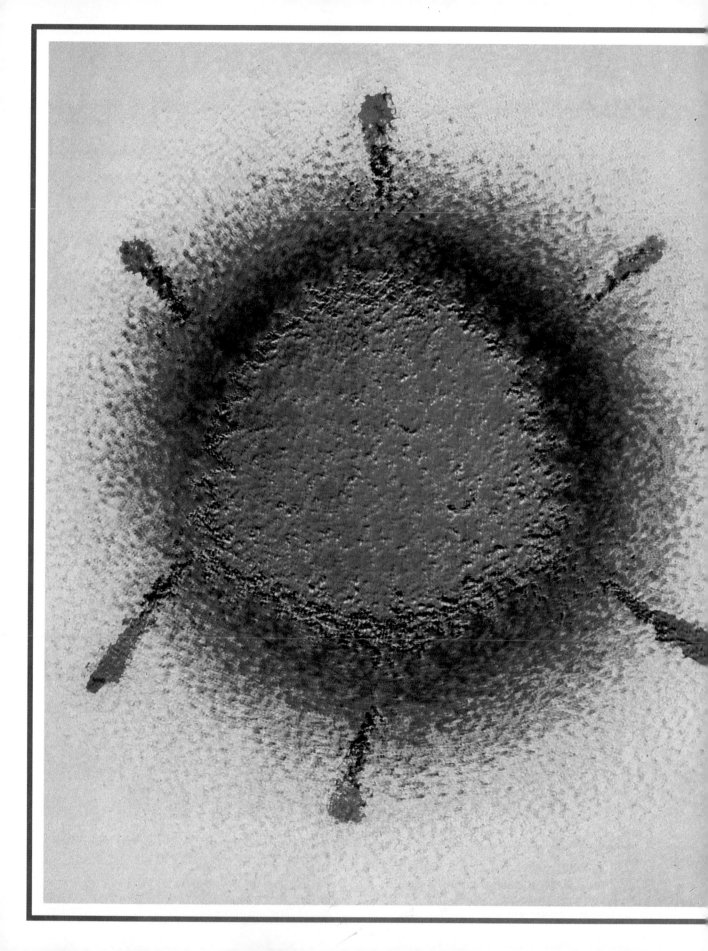

Viruses and Monerans

Guide for Reading

After you read the following sections, you will be able to

2–1 Viruses

- List the parts of a virus.
- Describe how a virus reproduces and causes disease.

2–2 Monerans

- Name and describe the parts of a moneran.
- Compare autotrophic and heterotrophic monerans.
- Discuss the helpful and harmful effects of the monerans.

Achoo! And so it begins—the misery of the most common illness—the cold. Colds can be caused by any one of more than 100 kinds of viruses. One such culprit is shown on the opposite page. Because you can be infected by one kind of cold virus one month and another kind of cold virus the next month, you can get several colds every year.

Tired of catching colds? Then take heart, for help is on the way. In 1986, scientists working at the University of Virginia School of Medicine were successful in stopping the attacks of some common cold viruses. The scientists are now seeking a way to stop the cold virus before it can cause an infection. To do so, they have developed disease-fighting substances called antibodies in the laboratory. These antibodies keep the cold virus from invading unsuspecting cells. By acting like a protective shield, they guard the body from invaders.

As you read this chapter, you will learn a great deal about viruses. You will also find out about tiny one-celled organisms called monerans. So turn the page and begin your journey into a strange world of incredibly small but fascinating living things.

Journal *Activity*

You and Your World In your journal, make a list of ten questions you have about diseases. As you study this chapter, see how many of your questions are answered. Devise a plan for finding out the answers to any questions that remain unanswered.

This virus causes one kind of cold. The viruses that cause colds are remarkably small—often little more than 10 billionths of a meter across.

2–1 Viruses

Imagine for a moment that you have been presented with a rather serious problem. A disease is killing one of your country's most important crops. Sick plants develop a pattern of yellow spots on their leaves. Eventually, the leaves wither and fall off, and the plant dies. You must find out what is causing the disease. With this information, it may be possible to save the remaining plants.

You gather some leaves from the sick plants and crush the leaves until they produce a juice. Then you put a few drops of the juice on the leaves of healthy plants. Several days later, you discover yellow spots on the once-healthy leaves. You reason that the cause of the disease can be found in the juice from the sick plants.

You then put the juice through a filter whose holes are so small that not even cells can slip through. You figure that this should take out any disease-causing microorganisms (microscopic organisms) in the juice. In fact, when you use the best light microscope to examine the filtered juice, you find no trace of microorganisms. But to your surprise, the juice still causes the disease in healthy plants! You realize that you must have discovered a germ that is smaller than a cell—too small to be

Figure 2–1 *Viruses come in many shapes. Rabies virus is shaped like a thimble (bottom left). Tobacco mosaic virus is rod-shaped (right). Rubella (German measles) virus is spherical (top left). The viruses that cause chicken pox and related diseases are oval (center) and are the largest viruses—as much as several hundred billionths of a meter long.*

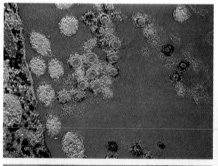

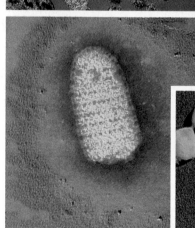

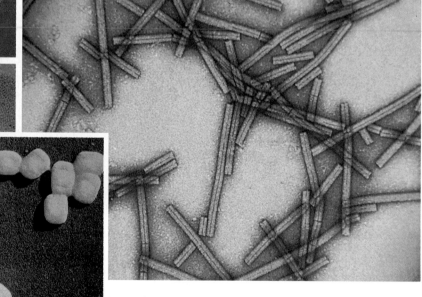

seen even with a microscope. But how could this be, you wonder. Cells are the basic unit of structure and function in living things. All living things are made of cells. There cannot be a living thing smaller than a cell! Or can there?

An additional experiment shows that the disease-causing germ can reproduce in the newly infected plants. This seems to indicate that a living thing is causing the disease. After all, the ability to reproduce is a characteristic of living things. A non-living, sickness-causing substance—a poison, for example—is not able to reproduce itself.

However, the results of other experiments seem to contradict the hypothesis that the disease-causing germ is alive. You discover that the crop-killing germ cannot be grown outside of the plants it infects. This is strange, because all living things can grow by themselves (provided they are given the correct nutrients and environmental conditions, of course). What is going on here, you wonder.

Finally, you decide to use chemical techniques to isolate the germ. After you purify enormous quantities of the juice from the infected plants, you are left with a tiny amount of whitish, needlelike crystals. These crystals show no evidence of being alive. They do not grow, breathe, eat, reproduce, or perform any other life functions. But when you inject the seemingly lifeless crystals into a healthy plant, the plant develops the disease. Clearly the mysterious crystals are the disease-causing germ itself. But what is this germ? And is it alive—or not?

This story is based on real events that started about 100 years ago and unfolded over the next forty years. The disease-causing germ is a **virus.** And whether viruses are alive or not depends on your definition of life. As you read in the story, there are good reasons to think of viruses as living things. And there are equally good reasons to think of viruses as nonliving things.

What Is a Virus?

Viruses are tiny particles that can invade living cells. Because viruses are not cells, they cannot perform all the functions of living cells. For example, they cannot take in food or get rid of wastes. In fact,

Figure 2–2 *Viruses cause a number of diseases in living things. The cherry's leaves are turning yellow and falling off as a result of a virus disease. Why are viruses considered parasites?*

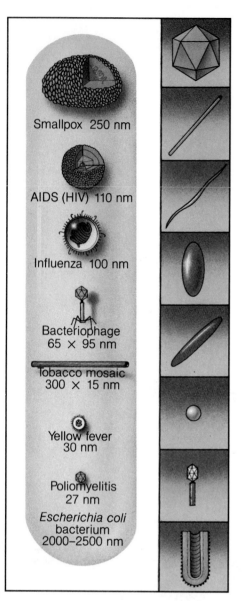

Figure 2–3 *Viruses have a wide variety of shapes and sizes. Viruses are measured in nanometers (nm). A nanometer is one-billionth of a meter. The yellow-green capsule-shaped structure represents a bacterial cell. In general, how do viruses compare to cells in terms of size? In terms of structure?*

Smallpox 250 nm

AIDS (HIV) 110 nm

Influenza 100 nm

Bacteriophage
65 × 95 nm

Tobacco mosaic
300 × 15 nm

Yellow fever
30 nm

Poliomyelitis
27 nm

Escherichia coli
bacterium
2000–2500 nm

Activity Bank

A Germ That Infects Germs,
p.171

about the only life function that viruses share with cells is reproduction. However, viruses cannot reproduce on their own. They need the help of living cells. The living cells are called **hosts.** Hosts are living things that provide a home and/or food for a **parasite** (PAIR-ah-sight). A parasite is an organism that survives by living on or in a host organism, thus harming it. Because viruses harm their host cells, they are considered to be parasites.

All five kingdoms of living things—plants, animals, fungi, protists, and monerans—are affected by viruses. In fact, experts suspect that all cells are subject to invasion by some kind of virus. It is interesting to note that each type of virus can infect only a few specific kinds of cells. For example, the rabies virus infects only nerve cells in the brain and spinal cord of dogs, humans, and other mammals. So a mammal's skin cells cannot be infected with rabies. And an organism that is not a mammal—such as a frog, plant, mushroom, or protist—cannot get rabies.

The origin of viruses is unknown. Because viruses need living cells in order to reproduce, it is likely that they appeared after the first cells. Many scientists think that viruses evolved from bits of hereditary material that were lost from host cells. This may mean that a virus is more closely related to its host than it is to other viruses. Thus the virus that causes your cold may be more closely related to you than it is to the virus that causes a plant's leaves to fall off!

Structure of Viruses

A virus has two basic parts: a core of hereditary material and an outer coat of protein. The hereditary material controls the production of new viruses. Like a turtle's shell, the protein coat encloses and protects the virus. The protein coat is so protective that some viruses survive after being dried and frozen for years. The protein coat also enables a virus to identify and attach to its host cell.

With the invention of the electron microscope in the 1930s, scientists were able to see and study the shapes and sizes of certain viruses. Some viruses, such as those that cause the common cold, have 20 surfaces. Each surface is in the shape of a triangle that has equal sides. Other viruses look like fine threads. Still others resemble tiny spheres. There are even some that look like miniature spaceships.

Reproduction of Viruses

In order to understand how a virus reproduces and causes disease, it might be helpful to examine the activities of one kind of virus known as a bacteriophage (bak-TEER-ee-oh-fayj). A bacteriophage is a virus that infects bacteria (singular: bacterium). In fact, the word bacteriophage means "bacteria eater." Bacteria are unicellular (one-celled) microorganisms that belong to the kingdom Monera.

In Figure 2–4, you can see how a bacteriophage (virus) attaches its tail to the outside of a bacterium. The bacteriophage quickly injects its hereditary material directly into the living cell. The protein coat is left behind. Once inside the cell, the bacteriophage's hereditary material takes control of all of the bacterium's activities. As a result, the bacterium is no longer in control. The bacterium begins to produce new bacteriophages rather than its own chemicals.

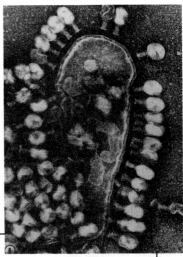

Figure 2–4 *The electron micrograph shows a bacterium under attack by numerous bacteriophages (inset). What stage in the diagram does this represent? What are the events in the life cycle of a bacteriophage?*

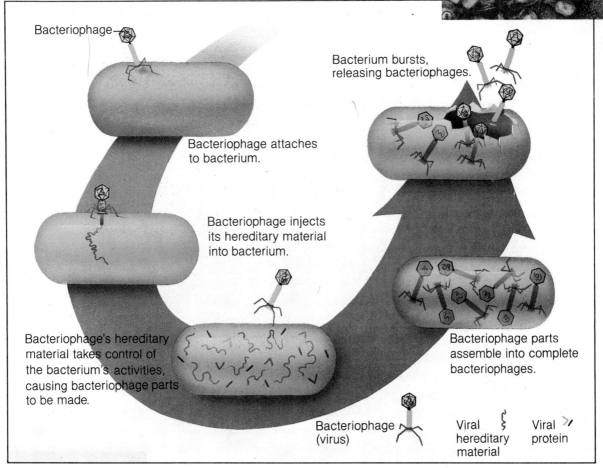

Bacteriophage

Bacteriophage attaches to bacterium.

Bacteriophage injects its hereditary material into bacterium.

Bacteriophage's hereditary material takes control of the bacterium's activities, causing bacteriophage parts to be made.

Bacterium bursts, releasing bacteriophages.

Bacteriophage parts assemble into complete bacteriophages.

Bacteriophage (virus)

Viral hereditary material

Viral protein

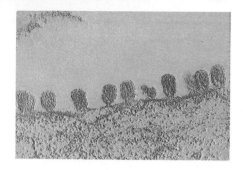

Figure 2–5 *Like many animal viruses, influenza viruses escape from their host cell by forming tiny bubbles at its surface.*

Figure 2–6 *A bacteriophage looks like a miniature spaceship. What are the main parts of a bacteriophage? What kind of cell is host to a bacteriophage?*

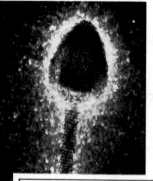

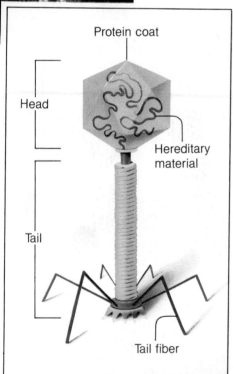

Protein coat

Head

Hereditary material

Tail

Tail fiber

Soon the bacterium fills up with new bacteriophages, perhaps as many as several hundred. Eventually, the bacterium bursts open. The new bacteriophages are released and infect nearby bacteria.

Not all viruses act like a bacteriophage. Some viruses keep their protein coat when they enter their host cell. Others, including some bacteriophages, simply join their hereditary material to that of the host cell and cause no immediate effects. A few cause disease by changing the behavior of their host cell and making it into a cancer cell. And many viruses do not cause their host cell to burst. Instead, a newly made virus particle causes the host cell to produce a tiny bubblelike structure on its edge. As you can see in Figure 2–5, this bubble eventually pinches off the side of the cell, carrying the virus with it.

Although the details of the "life" cycle vary from virus to virus, the basic pattern is the same for all viruses. **First, a virus gets its hereditary material into the host cell. Then the host cell makes more virus particles. Finally, the virus particles leave the original host cell and infect new hosts.**

Viruses and Humans

Viruses cause a large number of human diseases. Some of these diseases—such as colds, fever blisters, and warts—are simply annoying and perhaps a bit painful. Others are serious and can cause permanent damage or even death. Among the diseases caused by viruses are AIDS, measles, influenza, hepatitis, smallpox, polio, encephalitis, mumps, and herpes.

Much of the research on viruses has concentrated on ways of preventing and treating viral diseases. However, researchers have found ways of using viruses to help humans. Weakened or killed viruses are used to make some vaccines. A vaccine stimulates the body to produce antibodies (substances that prevent infection). Viruses can also be used to wage germ warfare on disease-causing bacteria and insects and on other agricultural pests. In the 1950s, a virus was used to control the population of rabbits in Australia. The virus killed about 80 percent of the

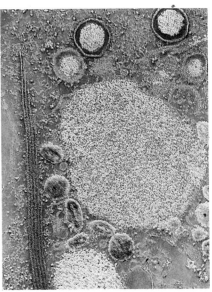

rabbits, enabling people to reclaim land for livestock and wildlife. Why might viruses be a better weapon against pests than poisons?

Recently, researchers have learned to put hereditary material into viruses. They can then use the viruses to put the hereditary material into cells. In the not-too-distant future, scientists may be able to use viruses to replace faulty information in a person's hereditary material. This could cure diseases such as diabetes, cystic fibrosis, sickle cell anemia, and many other hereditary disorders. Scientists may also be able to use viruses to improve crops. For example, corn plants might be "infected" with hereditary material that enables them to make their own fertilizer or resist pests.

Figure 2–7 *The population of rabbits in Australia was brought under control with the help of the virus that causes the rabbit disease myxomatosis. Why are viruses good for controlling pests? Can you think of some situations in which using viruses for pest control might be a bad idea?*

Figure 2–8 *The red circles are herpes viruses inside a human cell. The colors that you see do not actually exist in real life. They are added to photographs to make structures easier to see.*

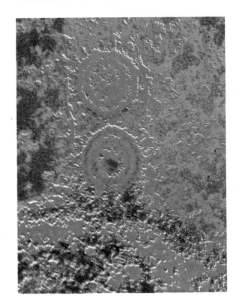

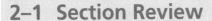

2–1 Section Review

1. How does a virus reproduce? How does this relate to how the virus causes disease?
2. Would you classify viruses as living or nonliving? Explain.
3. What is a bacteriophage?
4. Describe the structure of a virus.

Connection—*Technology*
5. Why were scientists unable to study the structure of viruses until after the electron microscope was invented?

CONNECTIONS

Computer Viruses: An Electronic Epidemic

"Shuttle to mission control. Do you read me?"

"We read you loud and clear. Commence data transmission." The computer screen flickered with changing numbers and words as mission control began to receive astronomical information from the Space Shuttle. Suddenly, the screen went blank.

Before the startled technician could adjust the computer's controls, a familiar television character appeared where the data should have been.

"Cookie!" it demanded. The technician stared at the screen, horrified. She then tried to type instructions. But the computer did not respond.

"No cookie, no play. Bye-bye." With these words, the figure on the screen disappeared. And so did all the hard-won information from the Space Shuttle. Within only a few seconds, a computer virus had undone years of work!

Although all the events in this story are imaginary, they are not impossible. Because many activities of modern life involve computers,

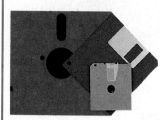

computer viruses are a matter of great concern. A computer virus is not really a virus. It is simply a program, or set of instructions to a computer. Computer viruses do not invade cells or cause diseases. They do not harm living things—at least not directly. They are made by humans and are definitely not alive. And the information they contain is in the form of magnetic or electric signals.

Some computer viruses, such as the one in the story you just read, are carried via telephone wires. Others are transmitted by computer diskettes that are "infected" with the virus program. If such diskettes are used, the virus is stored in the computer. The virus program then causes the computer to copy the computer virus onto every diskette it contacts in the future!

Don't be a victim of a virus! Whether biological or computer, viruses are enough to make you sick. A word to the wise is sufficient: Protect yourself from getting viruses and spreading them!

2–2 Monerans

Monerans are tiny organisms that consist of a single cell. As you learned in Chapter 1, moneran cells are different from all other cells because they lack a nucleus and certain other cell structures. At one time, the term bacteria was used to refer to only certain kinds of monerans. Other monerans were known as blue-green algae. Blue-green algae are now known as cyanobacteria, or blue-green bacteria. (The prefix *cyano-* means blue.) Because all monerans are now considered to be bacteria, the terms bacteria and monerans are used interchangeably. We will usually use the term bacteria in this chapter because it is more familiar and used more often than the term monerans. As you read this chapter, keep in mind that monerans and bacteria are the same thing.

As you may recall from Chapter 1, bacteria are the oldest forms of life on Earth. The first bacteria appeared about 3.5 billion years ago. Bacteria were Earth's only living things for about 2 billion years.

Bacteria are among the most numerous organisms on Earth. Scientists estimate that there are about 2.5 billion bacteria in a gram of garden soil. And the total number of bacteria living in your mouth is greater than the number of people who have ever lived!

As you might expect of such a large, ancient group, bacteria are quite varied. They may be rod-shaped like a medicine capsule, round as a marble, coiled like a stretched spring, round and stalked like a candied apple on a stick, or completely shapeless. They come in colors ranging from reds and yellows

Guide for Reading

Focus on this question as you read.

▶ How do monerans fit into the world?

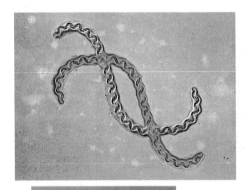

Figure 2–9 *The most common shapes of bacteria are spheres, rods, and spirals. However, some bacteria—such as Y-shaped Bifidiobacterium—have unusual shapes that do not fit into any of these categories.*

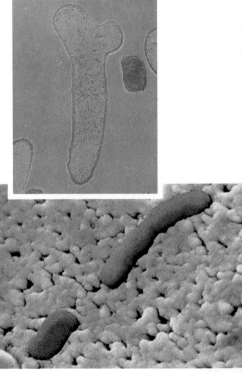

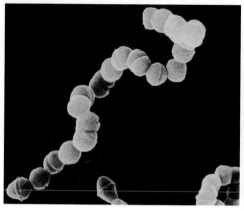

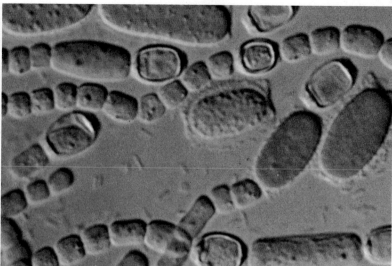

Figure 2–10 *Bacteria, such as many blue-green bacteria (right), may live in groups of cells attached to one another. The name of a bacterium can be a clue to what it looks like. For example, strepto- means chain and -coccus means a spherical bacterium. Why was this bacterium named Streptococcus (left)?*

to blues and violets. Some bacteria live alone as single cells. Others live in groups of cells that are attached to one another.

Bacteria are found in water, air, soil, and the bodies of larger organisms. In fact, bacteria live almost everywhere—even in places where other living things cannot survive. For example, some bacteria live in volcanic vents at the bottom of the ocean. The temperature of the water in these vents can be as high as 250°C—two and one-half times the temperature of boiling water.

Bacteria are considered the simplest organisms. However, bacteria are more complex than they may appear. Each bacterial cell performs the same basic functions that more complex organisms, including you, perform.

Structure of Bacteria

One of the most noticeable features of a bacterium is the cell wall. See Figure 2–12. The cell wall is a tough, rigid structure that surrounds, supports, shapes, and protects the cell. Almost all bacteria have a cell wall. In some bacteria, there is a coating on the outside of the cell wall. This coating is called the capsule. How might the capsule provide protection for a bacterium?

Lining the inside of the cell wall is the cell membrane. The cell membrane controls which substances enter and leave the cell. Within the cell membrane

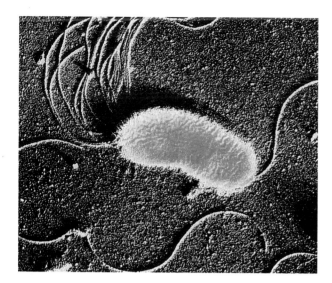

Figure 2–11 *The long, thin, whiplike structures on this soil bacterium are flagella. What is the function of flagella?*

is the cytoplasm. The cytoplasm is a jellylike mixture of substances that makes up most of the cell.

Unlike most other cells, the hereditary material of bacteria is not confined in a nucleus. (A nucleus is a membrane-enclosed structure that can be thought of as the "control center" of a typical cell.) In other words, there is no membrane separating the hereditary material from the rest of the cell in monerans.

Many bacteria are not able to move on their own. They can be carried from one place to another by air and water currents, clothing, and other objects. Other bacteria have special structures that help them move in watery surroundings. One such structure is a flagellum (flah-JEHL-uhm; plural: flagella). A flagellum is a long, thin, whiplike structure that propels a bacterium through its environment. Some bacteria may have many flagella.

Life Functions of Bacteria

Bacteria have more different ways of getting the energy they need to live than any other kingdom of organisms. In fact, bacteria obtain energy in more ways than all of the other kingdoms combined. Like most other organisms, many bacteria need oxygen in order to get energy from food. Other bacteria can thrive without oxygen. And still other kinds of bacteria will die if they are exposed to oxygen.

Many bacteria are heterotrophs. Recall from Chapter 1 that a heterotroph cannot make its own food. A heterotroph gets energy by eating food,

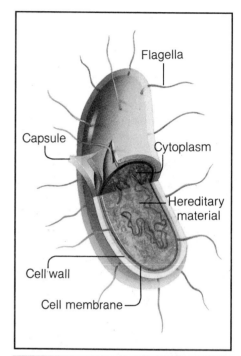

Figure 2–11 *The long, thin, whiplike structures on this soil bacterium are flagella. What is the function of flagella?*

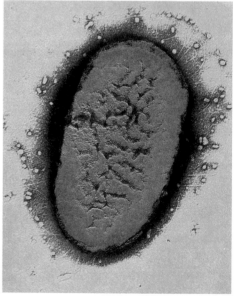

Figure 2–12 *The diagram shows the structure of a typical bacterium. Can you locate the flagella, capsule, cytoplasm, and genetic material on the photograph of the whooping cough bacterium?*

Figure 2–13 *If you could look at this scene with a powerful microscope, you would discover bacteria busily breaking down the remains of the dead tree. Why are some bacteria called decomposers?*

usually other organisms. Some bacteria feed on living organisms. These bacteria are parasites. As you just learned, parasites are organisms that live and feed either inside or attached to the outer surface of a host organism, thus harming the host. Such bacteria cause infections in people, animals, and plants. Other bacteria feed on dead things. These bacteria are **decomposers.** Decomposers break down dead organisms into simpler substances. In the process, they return important materials to the soil and water.

Some bacteria are autotrophs. An autotroph, you will recall, makes its own food. Some food-making bacteria use the energy of sunlight to produce food. Other bacteria use the energy in certain substances that contain sulfur and iron to make food. The nauseating "sulfur" smell of mud flats or rotting food is due to the action of such bacteria.

When food is plentiful and the environment is favorable, bacterial cells grow and then reproduce by dividing into two cells. Under the best conditions, most bacteria reproduce quickly. Some types can double in number every 20 minutes. At this rate, after about 24 hours the offspring of a single bacterium would have a mass greater than 2 million kilograms, or as much as 2000 mid-sized cars! In a few more days, their mass would be greater than that of the Earth. Obviously, this does not happen. Why do you think this is so?

When food is scarce or conditions become unfavorable in other ways, some bacteria form a small internal resting cell called an endospore. As you can see in Figure 2–15, an endospore consists of hereditary material, a small amount of cytoplasm, and a

Figure 2–14 *Bacteria reproduce by splitting into two cells (left). This process increases the number of bacteria. There are also processes that produce new bacteria but do not increase the number. An old bacterium is changed into a new one when genetic material is transferred via a special tube from one bacterium to another (right).*

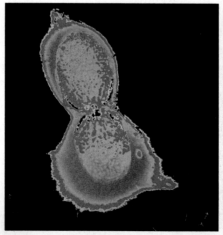
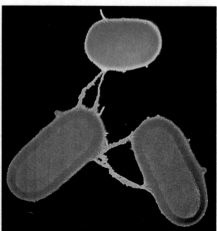

thick protective outer coat. An endospore can survive long periods in which the environment is not suitable for bacterial growth. Some endospores can survive being touched by disinfectant chemicals, blown through the atmosphere, frozen in polar ice, baked in the desert sun, boiled for an hour, and bombarded with powerful radiation. When environmental conditions improve, the endospores burst out of their protective coats and develop into active bacteria. Can you now explain why bacteria are found all over the world—and why some bacteria are extremely hard to kill?

Bacteria in Nature

Most types of bacteria are not harmful and do not cause disease. In fact, many types of bacteria are helpful to other living things and perform important jobs in the natural world.

FOOD AND ENERGY RELATIONSHIPS **Bacteria are an essential part of the food and energy relationships that link all life on Earth.** As you learned earlier, some bacteria (decomposers) break down dead materials to form simpler substances. These simpler substances can be used by autotrophs—such as green plants and blue-green bacteria—to make food. Small heterotrophs (organisms that cannot make their own food) such as certain protists and tiny animals feed on plants and blue-green bacteria. These small heterotrophs are food for large heterotrophs, which are

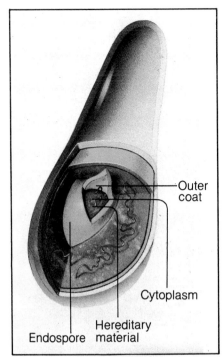

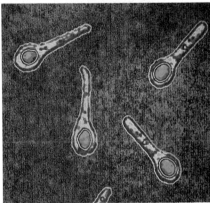

Figure 2–15 *The bacterium that causes the disease tetanus, or lockjaw, forms endospores. The structure of an endospore is shown in the diagram. What is the function of an endospore?*

Figure 2–16 *Flamingoes feed on blue-green bacteria and small organisms, which they filter from the water with their beaks. The structure of their filtering beak causes flamingoes to do "headstands" when feeding. Colored substances in the blue-green bacteria give flamingoes their pink color.*

ACTIVITY

in turn food for still larger heterotrophs. When heterotrophs and autotrophs die, they become food for decomposers, such as certain bacteria. Thus the cycle continues.

OXYGEN PRODUCTION   The food-making process of blue-green bacteria produces oxygen as well as food. Billions of years ago, this resulted in a dramatic change in the composition of the Earth's atmosphere. The percentage of oxygen gas increased from less than 1 percent to about 20 percent. This made it possible for oxygen-using organisms such as protists, plants, animals, and fungi to evolve.

CHANGING ENVIRONMENTS   Bacteria continue to change environments in ways that make them suitable for other organisms. For example, bacteria are among the first organisms to grow on the bare rock created by the action of volcanoes. As they grow, the bacteria break down the rock and help create soil. Soon small plants can take root in the developing soil. The bacteria and small plants provide food and homes for tiny animals, fungi, and protists. These new arrivals further change the environment, making it possible for larger organisms to survive. Eventually, a variety of large and small organisms live on what was once bare, lifeless rock.

SYMBIOSES   Some bacteria help other organisms by forming a partnership, or **symbiosis**· (sihm-bigh-OH-sihs; plural: symbioses). Symbiosis is a relationship in which one organism lives on, near, or even inside another organism and at least one of the organisms benefits. Here is an example. Food-making

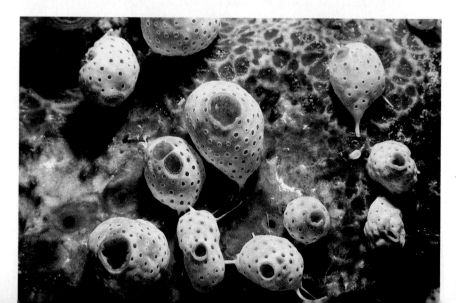

Figure 2–17 *Certain sea squirts are involved in a symbiosis with food-making bacteria. What is symbiosis?*

bacteria live in the bodies of sea squirts and deep-sea tube worms. The bacteria are protected by the body of their host. In return, they provide their host with food. In another example of symbiosis, certain bacteria that live in the intestines of animals such as cows, termites, horses, and humans break down plant cell walls. This enables the host to digest plant materials such as wood, grass, and fruit. Bacteria called nitrogen-fixing bacteria turn nitrogen gas, which plants cannot use, into nitrogen compounds that plants can use to make biologically important substances. Nitrogen-fixing bacteria also help replace the nitrogen compounds in the soil. Without such nitrogen-fixing bacteria, most nitrogen compounds in the soil would be quickly used up and plants could no longer grow. Some nitrogen-fixing bacteria live as individual cells in soil. Others form strandlike colonies in water. Still others live in lumps on the roots of plants such as alfalfa, soybean, and clover. See Figure 2–18.

Bacteria and Humans

Bacteria help humans in many ways. Bacteria are involved in the production of food, fuel, medicines, and other useful products. Some are used in industrial processes. Others help break down pollutants, which are substances such as waste materials or harmful chemicals that dirty the environment.

Although most bacteria are either helpful or harmless, a few can cause trouble for humans. The trouble comes in a number of forms. Some harmful bacteria spoil food or poison water supplies. Others damage property or disrupt manufacturing processes. Still others cause diseases in people, pets, livestock, and food crops.

FOOD Many food products, especially dairy products, are produced with the help of bacteria. Bacteria (and products made by bacteria) are used to make cheeses, butter, yogurt, sour cream, pickles, soy sauce, vinegar, and high-fructose corn syrup. Nitrogen-fixing bacteria provide substances that crops need to grow. In flooded fields such as those used to grow rice, nitrogen-fixing blue-green bacteria fertilize the crops naturally. But bacteria do more

Figure 2–18 *The nodules, or lumps, on the roots of the pea plant are home to symbiotic nitrogen-fixing bacteria (top). If you were to break open a nodule such as this, you would discover the bacteria housed inside (bottom). How does this symbiosis help the pea plant and the bacteria?*

Activity Bank

Yuck! What Are Those Bacteria Doing in My Yogurt? p. 172

Figure 2–19 *Some bacteria harm humans and livestock when they grow in and poison water supplies.*

than help to make foods. They can serve as the food itself. Blue-green bacteria, which may grow in water as large masses of strands, have long been used as food. True, blue-green bacteria might not sound very appetizing, but they are quite nutritious. They are about 70 percent protein, rich in vitamins and other nutrients, and easy to digest.

Some helpful bacteria break down food to make tasty or useful products. For example, one kind of helpful bacterium breaks down milk to produce yogurt. Unfortunately, harmful bacteria may also break down food, making smelly, bad-tasting, or even poisonous products. In other words, some harmful bacteria may cause food to spoil.

Food spoilage can be prevented or slowed down by heating, drying, salting, cooling, or smoking foods. Each of these processes prevents or slows down the growth of bacteria. For example, milk is heated to 71°C for 15 seconds before it is placed in containers and shipped to the grocery or supermarket. This process, called pasteurization, destroys most of the bacteria that would cause the milk to spoil quickly. Heating and then canning foods such as vegetables, fruits, meat, and fish are also used to prevent bacterial growth. But if the foods are not sufficiently heated before canning, bacteria can grow inside the can and produce poisons called toxins. Toxin-producing bacteria may also produce a gas

Figure 2–20 *Although the 1990 oil spill in Huntington Beach, California, was not particularly large, it still took a great deal of effort to clean it up. Oil-eating bacteria are now being used on an experimental basis to help clean up such spills.*

that causes the can to bulge. You should never eat food from such a can.

Harmful bacteria can also spoil water supplies. When environmental conditions are changed by pollution or other causes, blue-green bacteria and other bacteria in a body of water may multiply rapidly. The huge numbers of bacteria form an ugly, often smelly, scum and poison the water. Humans and livestock can get very sick from drinking the poisoned water.

FUEL Certain bacteria break down garbage such as fruit rinds, dead plants, manure, and sewage to produce methane. Methane is a natural gas that can be used for cooking and heating. Over millions and millions of years, heat and pressure within the Earth changed the remains of ancient blue-green bacteria, among other organisms, into an oily mixture of chemicals. This mixture of chemicals is called petroleum. Petroleum is the source of heating oil, gasoline, kerosene, and many other useful substances.

ENVIRONMENTAL CLEANUP Bacteria clean up the environment in many ways. Some are used to treat sewage; others cause garbage to decompose, or rot. A few types of bacteria are able to break down the oil in oil spills. Still others break down complex chemicals such as certain pesticides and a few types of plastic.

HEALTH AND MEDICINE Some bacteria help to keep you healthy. For example, bacteria that live inside

ACTIVITY

Bacterial Growth

If a bacterium reproduces every 20 minutes, how many bacteria would there be after one hour? After two hours? After four hours? After eight hours? (Assume all bacteria survive and reproduce.)

If each bacterium is 0.005 mm long, how many bacteria, laid end to end, would equal 1 mm in length?

Using a metric ruler, determine the length of your thumbnail in millimeters. How many 0.005 mm-long bacteria laid end to end would equal the length of your thumbnail?

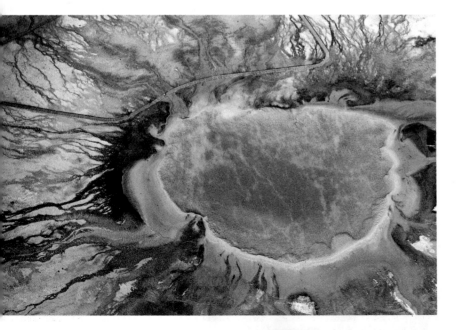

Figure 2–21 *Ancient blue-green bacteria, similar to the modern ones that give the hot springs in Yellowstone National Park their beautiful colors, were the source of the fuel called petroleum.*

the human intestines help you to digest food. They also make vitamins that the body cannot make on its own. And they may help to prevent disease by crowding out harmful bacteria.

Certain helpful bacteria produce substances that are used to fight disease. Some bacteria naturally produce chemicals that have been found to destroy or weaken other bacteria. Such chemicals are known as **antibiotics.** Why would the production of antibiotics be useful to such bacteria? Recently, scientists have developed ways to change the hereditary material in bacteria. By putting new information into the bacteria's hereditary material, scientists cause the bacteria to produce medicines and other useful substances.

Some human diseases caused by bacteria include strep throat, certain kinds of pneumonia, diphtheria, cholera, tetanus, tuberculosis, bubonic plague, and Lyme disease. Bacteria that live in the mouth cause tooth decay and gum disease. Some of these diseases can be prevented by proper hygiene or immunization shots. Others can be treated with antibiotics.

INDUSTRY   Helpful bacteria have long been used in the process of tanning leather. Recently, people have begun to use bacteria to extract copper, gold, and other useful and valuable metals from rock. Interestingly, some types of metal-rich rocks, or ores, may have been created by the actions of bacteria that lived millions of years ago. Scientists hope to develop an inexpensive way to use bacteria to make plastics and other complex compounds.

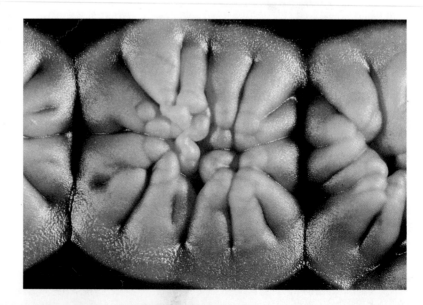

Figure 2–22 *The lumpy growths of* Streptomyces *are made of huge numbers of bacteria growing together.* Streptomyces *is the source of many antibiotics. Why are antibiotics useful to humans?*

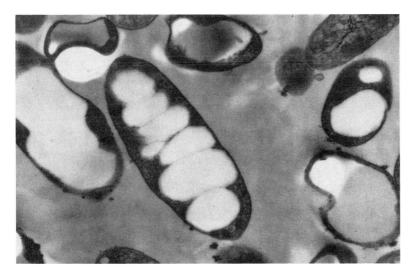

Although the time is not yet right for bacterial plastics, other substances made by bacteria are manufactured commercially. These substances are used for many different purposes—coloring food and cosmetics, tenderizing meat, removing stains, processing paper and cloth, and changing one chemical to another, to name a few.

Harmful bacteria can disrupt industrial processes. They damage leather during the tanning process, ruin paper pulp, and turn fruit juice and wine into vinegar. Some harmful bacteria break down asphalt and thus damage roads, parking lots, and other paved surfaces. And bacteria cause millions of dollars of damage each year to oil-drilling machinery, water and gas pipes, and supplies of petroleum.

ACTIVITY

Benefiting From Bacteria

Write a 200-word essay on some of the useful ways bacteria contribute to nature and humankind. Give specific examples.

2–2 Section Review

1. Describe two ways in which bacteria are helpful and two ways in which they are harmful.
2. What is an endospore? How does it help bacteria survive?
3. Describe the structure of a bacterium.

Critical Thinking—*Applying Concepts*
4. Many antibiotics work by damaging a bacterium's cell wall. Explain why such antibiotics are not effective against viruses and certain kinds of bacteria that lack cell walls.

Laboratory Investigation

Examining Bacteria

Problem

Where are bacteria (monerans) found?

Materials (per group)

5 petri dishes with sterile nutrient agar
glass-marking pencil
pencil with eraser
soap and water

Procedure 🔬

1. Turn each petri dish bottom side up on the table. **Note:** *Be careful not to open the petri dish.*

2. With the glass-marking pencil, label the bottom of the petri dishes containing the sterile nutrient agar A, B, C, D, and E. Turn the petri dishes right side up.

3. Remove the lid of dish A and lightly rub a pencil eraser across the agar in the petri dish. Close the dish immediately.

4. Remove the lid of dish B and leave it open to the air until the class period ends. Then close the lid.

5. Remove the lid of dish C and lightly run your index finger over the surface of the agar. Then close the lid immediately.

6. Wash your hands thoroughly. Remove the lid of dish D and lightly rub the same index finger over the surface of the agar. Then close the lid immediately.

7. Do not open dish E.

8. Place all five dishes upside down in a warm, dark place for three or four days.

9. After three or four days, examine each dish. **CAUTION:** *Do not open dishes.*

10. On a sheet of paper, construct a table similar to the one shown here. Fill in the table.

11. Return the petri dishes to your teacher. Your teacher will properly dispose of them.

Observations

1. How many clusters of bacteria appear to be growing on each petri dish? Are there different types of clusters?

2. Which petri dish has the most bacterial growth? Which has the least?

Petri Dish	Source	Description of Bacterial Colonies
A		
B		
C		
D		
E		

Analysis and Conclusions

1. Which petri dish was the control? Explain.

2. Did the dish that you touched with your unwashed finger contain more or less bacteria than the one that you touched with your washed finger? Explain.

3. Explain why the agar was sterilized before the investigation.

4. Design an experiment to show if a particular antibiotic will inhibit bacterial growth in a petri dish.

5. Suggest some methods that might stop the growth of bacteria.

6. What kinds of environmental conditions seem to influence where bacteria are found?

Summarizing Key Concepts

2–1 Viruses

▲ Viruses are tiny noncellular particles that infect living cells.

▲ Viruses are parasites that probably affect all types of cells. Each type of virus can infect only a few specific kinds of cells.

▲ Viruses cannot carry out any life functions unless they are in a host cell. Viruses may have evolved from bits of hereditary material that were lost from host cells.

▲ The development of the electron microscope made it possible to see viruses.

▲ Viruses consist of a core of hereditary material and an outer coat of protein.

▲ The basic "life" cycle is the same for all viruses. First, a virus gets its hereditary material into the host cell. Then the host cell makes more virus particles. Finally, the virus particles infect new hosts.

▲ Viruses cause a number of diseases that range from annoying to serious to fatal.

▲ Viruses cause human diseases such as colds, chicken pox, rabies, polio, and AIDS.

▲ Weakened or killed viruses are used to make some vaccines.

▲ Viruses are used to kill pests. They are also used to change the hereditary material of cells in specific ways.

2–2 Monerans

▲ Monerans are single-celled organisms that lack a nucleus and many other cell structures.

▲ The terms monerans and bacteria are interchangeable.

▲ Bacteria are the oldest forms of life on Earth. They are also among the most numerous and varied.

▲ Some bacteria live in colonies. A colony is a group of organisms that live together in close association. Some members of these colonies may be specialized for specific functions.

▲ Some bacteria are heterotrophs. Some heterotrophic bacteria are parasites. Others are decomposers.

▲ Some bacteria are autotrophs. Some autotrophic bacteria produce oxygen.

▲ Bacteria fit into the world in many ways. They are involved in many food and energy relationships with other organisms. Some change environments in ways that make them suitable for other organisms—by making soil, oxygen, or nitrogen compounds, for example. Others are involved in symbioses with other organisms.

▲ Some bacteria are helpful to humans. Others are harmful.

Reviewing Key Terms

Define each term in a complete sentence.

2–1 Viruses
   virus
   host
   parasite

2–2 Monerans
   decomposer
   symbiosis
   antibiotic

Chapter Review

Content Review

Multiple Choice

Choose the letter of the answer that best completes each statement.

1. An example of a disease caused by a virus is
   a. bubonic plague.   c. strep throat.
   b. measles.   d. tetanus.

2. Which of the following statements is true?
   a. Because they break down wastes and dead organisms, viruses are called decomposers.
   b. Unlike other cells, viruses lack a nucleus.
   c. Viruses consist of a core of protein surrounded by a coat of hereditary material.
   d. To perform their life functions, viruses require a host cell.

3. Almost all bacteria are surrounded and supported by a tough, rigid protective structure called the
   a. cell membrane. c. protein coat.
   b. cell wall.   d. capsule.

4. Which of the following is not found in bacteria?
   a. hereditary material   c. cytoplasm
   b. cell membrane   d. nucleus

5. Viruses are best described as
   a. autotrophs.   c. parasites.
   b. decomposers.   d. lithotrophs.

6. Bacteria cause a number of human diseases, including
   a. tuberculosis.   c. AIDS.
   b. influenza.   d. rabies.

7. Humans use viruses to
   a. make antibiotics.
   b. make fuels such as methane.
   c. break down sewage.
   d. put new hereditary material into cells.

8. Monerans are also known as
   a. bacteria.   c. viruses.
   b. bacteriophages.   d. fungi.

True or False

If the statement is true, write "true." If it is false, change the underlined word or words to make the statement true.

1. Organisms that break down wastes and the remains of dead plants and animals are called <u>producers</u>.

2. Some bacteria use whiplike structures called <u>flagella</u> to propel them through their environment.

3. When conditions become unfavorable, some bacteria produce a small internal resting cell called a <u>capsule</u>.

4. <u>Symbiosis</u> is a relationship in which one organism lives in close association with another organism and at least one of the organisms benefits.

5. <u>Bacteria</u> can be used by scientists to insert hereditary material into cells.

Concept Mapping

Complete the following concept map for Section 2–1. Refer to pages B6–B7 to construct a concept map for the entire chapter.

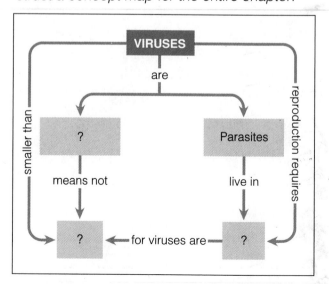

Concept Mastery

Discuss each of the following in a brief paragraph.

1. How are viruses different from cells?
2. Describe how a typical bacteriophage reproduces.
3. Identify three ways in which viruses help people.
4. How do autotrophs and heterotrophs differ in the ways they obtain food?
5. What is symbiosis? Give one example of symbiosis.
6. Briefly describe the structure of a typical bacterial cell. What is the most important structural difference between the cells of bacteria and the cells of organisms in the other four kingdoms?
7. How do bacteria affect food production, food processing, and food storage?
8. Can viruses be grown in the laboratory on synthetic material? Explain.

Critical Thinking and Problem Solving

Use the skills you have developed in this chapter to answer each of the following.

1. **Applying definitions** Why is the relationship between a parasite and a host considered to be a form of symbiosis?
2. **Making inferences** When a cell is placed in water that contains a large quantity of dissolved substances such as sugar or salt, the cell will shrivel up and die. Explain why food can be preserved by putting it into honey or brine (very salty water).
3. **Interpreting a graph** The accompanying graph shows the growth of bacteria. Describe the growth of the bacteria using the numbers given for each growth stage. Why do you think the growth leveled off in stage 3 and then fell in stage 4?

4. **Designing an experiment** Design an experiment to test the effect of temperature on the growth of bacteria. Be sure you have a control and a variable in your experiment. What results do you expect to get in this experiment? How can you apply these results in order to control food spoilage?
5. **Applying concepts** Every few years, a farmer plants a field with alfalfa or clover. At the end of the growing season, the farmer does not harvest these plants. Instead, the alfalfa or clover is plowed under, and the field is replanted with grain. Explain how the actions of bacteria account for the farmer's behavior.
6. **Using the writing process** In *The War of the Worlds,* a wonderful book by H. G. Wells, the Earth is invaded by Martians. Human weapons are useless against the invaders, and the Earth seems doomed. The Earth is saved, however, when the invaders die from diseases they contract here. Many other science-fiction writers have written stories that involve diseases. Now it is your turn. Write a science-fiction story in which a disease plays an important part.

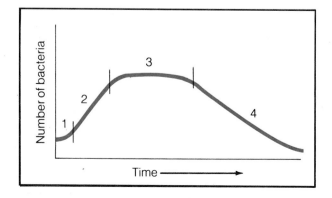

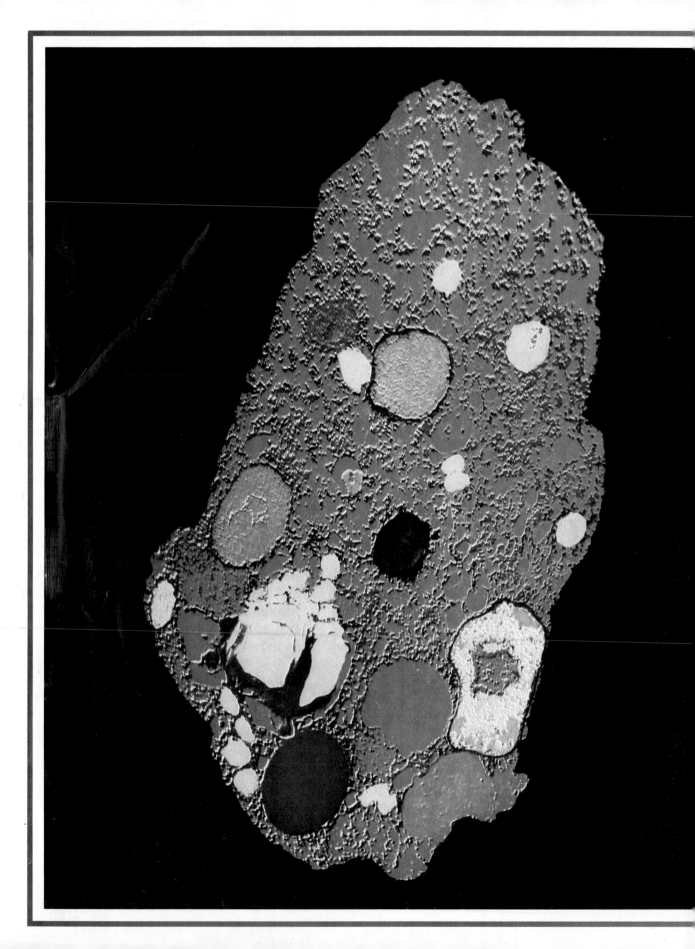

Protists

Guide for Reading

After you read the following sections, you will be able to

3–1 Characteristics of Protists
■ Describe the characteristics of protists.

3–2 Animallike Protists
■ Identify the four groups of animallike protists.

3–3 Plantlike Protists
■ Describe plantlike protists.

3–4 Funguslike Protists
■ Compare slime molds to other protists.

Imagine for a moment that you have stepped into a time machine. Your destination: the world of 1.5 billion years ago. You arrive in a strange and barren place. No animals roam the land or swim in the ocean. No birds fly in the air. No trees, shrubs, or grasses grow from the soil. Earth at first seems lifeless.

A closer look at your surroundings reveals a slippery black film on the rocks and greenish threads in the water. Bacteria live here! When you use a microscope to examine the water of early Earth, you discover that a new form of life has come into being. Like the bacteria that came before it, this form of life is unicellular (one-celled). But these cells are much more complex than those of bacteria. You are surprised and delighted to recognize one large structure in the cells—the nucleus.

These first cells with a nucleus are protists. Protists represent the first step in the series of evolutionary events that eventually led to the development of multicellular (many-celled) organisms such as mushrooms, trees, fishes, birds, and humans. In this chapter, you will learn a great deal about protists—complex and fascinating organisms.

Journal *Activity*

You and Your World When did you first find out that you were surrounded by living things too small to see? In your journal, explore the thoughts and feelings you had upon learning about the unseen world around you.

The numerous structures within this ameba look like tiny jewels. These structures represent an important evolutionary step toward increasing complexity in living things.

3–1 Characteristics of Protists

The members of the kingdom Protista are known as **protists.** Like bacteria (monerans), protists are unicellular (one-celled) organisms. Also like bacteria, protists were one of the first groups of living things to appear on Earth. Although protists are larger than monerans, most cannot be seen without the aid of a microscope. Protists, however, are quite different from bacteria. The most important difference is that protists have a nucleus and a number of other cell structures that bacteria lack.

Protists can be defined as single-celled organisms that contain a nucleus. However, you will discover in this chapter that there are a few protists that do not fit this definition perfectly. In fact, as scientists develop new techniques for examining the structure and chemistry of cells, the definition of protist continues to change. And as new evidence is obtained, analyzed, and interpreted, scientists revise their ideas about which protists should be classified together—and even about which organisms should be classified as protists!

Figure 3–1 *The major difference between protists and bacteria is that protists have many complex cell structures. The most important of these is the nucleus, which looks like a string of beads in* Stentor *(left), and a thin squiggle in* Vorticella *(center). Protists also have complex external structures, such as the hairlike projections on* Tetrahymena *(right).*

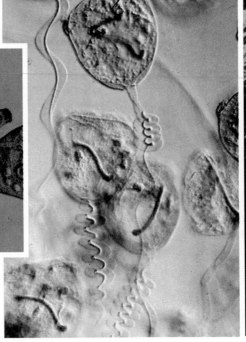

Most protists live in a watery environment. They can be found in the salty ocean and in bodies of fresh water. Some protists live in moist soil. Others live inside larger organisms. Many of these internal protists are parasites. (Parasites, you will recall, are organisms that live on or in a host organism and harm it.) But some of the protists that live inside other organisms help their hosts.

Protists generally live as individual cells. However, many protists live in colonies such as the ones shown in Figure 3–2. A protist colony consists of a number of relatively independent cells of the same species that are attached to one another. Some colonies are foamlike clusters of cells. Others are branching and treelike. A few are rings or chains of cells. And one type of colony consists of an intricate network of slime tunnels. The members of this type of colony move through the tunnels like miniature subway cars.

Evidence from fossils indicates that protists evolved about 1.5 billion years ago—about 2 billion years after bacteria. According to one hypothesis, the first protists were the result of an extremely successful symbiosis among several kinds of bacteria. Recall from Chapter 2 that a symbiosis is a close relationship of two or more organisms in which at least one organism benefits. In this case, some of the bacteria lived inside another bacterial cell. Each kind of bacterium performed activities that helped the "team" survive. Over time, the partners in the symbiosis became more specialized (suited for one particular task) and lost their independence. Traces of this early symbiosis can still be seen in protists and other cells that possess a nucleus. Within these cells are certain structures that are quite similar to bacterial cells. They even contain their own hereditary material.

Protists vary greatly in appearance and in the ways in which they carry out their life functions. For example, some protists are autotrophs. Autotrophs are able to use energy such as sunlight to make their own food from simple raw materials. Other protists are heterotrophs. Heterotrophs get their energy by eating food that has already been made. And a few protists are able to function as either autotrophs or heterotrophs, depending on their surroundings.

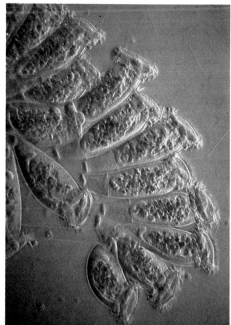
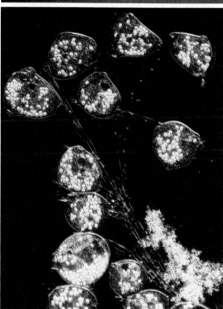

Figure 3–2 *Some protists, such as* Epistylis *(top) and* Carchesium *(bottom) live in colonies of attached but relatively independent cells. Colonial protists may have been the first evolutionary step toward true multicellular organisms.*

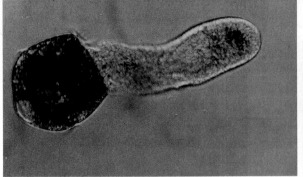

Figure 3–3 *These diatoms growing in fan-shaped colonies are autotrophs. The short, dark, thimble-shaped* Didinium *and the long, oval* Paramecium *it is eating àre both heterotrophs. What is the major difference between autotrophs and heterotrophs?*

As you might have guessed of such a large and diverse group of organisms, the kingdom Protista is divided into many phyla. Because there are many different ways to interpret what we know about protists, the classification of protists is a matter of scientific debate. Experts recognize anywhere from about nine to more than two dozen phyla. For the sake of simplicity, protists can also be grouped in three general categories. These categories, which are the ones you will read about here, are: animallike protists, plantlike protists, and funguslike protists.

3–1 Section Review

1. What are the major characteristics of protists?
2. Explain why scientists have not yet agreed on a single classification system for protists.
3. How are protists similar to monerans? How are they different?
4. When did protists first appear on Earth?

Connection—*You and Your World*

5. While standing in line at the supermarket, you notice the following newspaper headline: "Science Shocker: Alien Invaders in Every Cell of Your Body!" Explain why there may be some truth in this headline. (*Hint:* Fungi, plants, and animals evolved from protists that lived millions of years ago.)

3–2 Animallike Protists

Animallike protists are sometimes known as protozoa, which means first animals. Long ago, these organisms were classified as animals because they have several characteristics in common with animals. Their cells contain a nucleus and lack a cell wall. They are heterotrophs. Most of them can move. Animallike protists are no longer placed in a separate kingdom from more plantlike protists. Scientists have discovered that some animallike protists are so similar to certain plantlike protists that it does not make sense to place them in separate kingdoms.

Animallike protists are divided into four main groups. These four groups are the **sarcodines** (SAHR-koh-dighnz), the **ciliates** (SIHL-ee-ihts), the **zooflagellates** (zoh-oh-FLAJ-ehl-ihts), and the **sporozoans** (spohr-oh-ZOH-uhnz).

Sarcodines

Sarcodines are characterized by extensions of the cell membrane and cytoplasm known as pseudopods (SOO-doh-pahdz). The word **pseudopod** comes from the Greek words that mean "false foot" because these footlike extensions are always temporary. Pseudopods are used to capture and engulf particles of food. Some sarcodines also use psuedopods to move from one place to another.

Many sarcodines have shells that support and protect the cell. As you can see in Figure 3–4, these shells come in many forms. The shells of foraminiferans (fuh-ram-ih-NIHF-er-anz) may resemble coins, squiggly worms, the spiral burner coils of an electric stove, clusters of bubbles, and tiny sea shells. The shells of radiolarians (ray-dee-oh-LAIR-ee-uhnz) are studded with long spines and dotted with tiny holes. Because of this, radiolarian shells often look a lot like crystal holiday ornaments. When foraminiferans and radiolarians die, their shells sink to the ocean floor and form thick layers. Over millions of years, these shells are changed to rock. Some rocks that contain ancient protist shells, such as limestone and marble, are used in building. Others are used

Guide for Reading

Focus on this question as you read.

▶ *How are animallike protists similar? How are they different?*

Figure 3–4 *Many foraminiferans have beautiful shells. The shells of ancient foraminiferans help to make up limestone, marble, and chalk.*

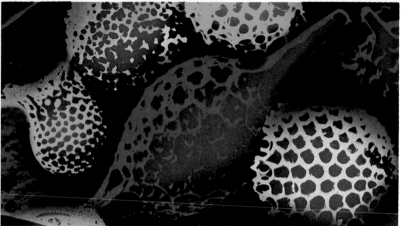

Figure 3–5 *The pores and spines of radiolarian shells make them glitter in the light like tiny glass decorations.*

Figure 3–6 *This diagram shows the structure of* Amoeba proteus, *a typical ameba. According to Greek mythology, the sea god Proteus had the magical ability to change his shape. How do amebas change their shape?*

for writing and drawing—the chalk used by teachers is made of prehistoric foraminiferans!

The most familiar type of sarcodine is the blob-like **ameba** (uh-MEE-bah). Amebas use their pseudopods to move and to obtain food. An ameba first extends a thick, round pseudopod from part of its cell. Then the rest of the cell flows into the pseudopod. As an ameba nears a small piece of food, such as a smaller protist, the ameba extends a pseudopod around the food. Soon the food particle is completely surrounded by the pseudopod. As you can see in Figure 3–8, this process produces a bubblelike structure that contains the food. This structure is called a food vacuole. The food is digested (broken down into simpler materials) inside the food vacuole. The digested food can then be used by the ameba for energy and growth. The waste products of digestion are eliminated when the food vacuole joins with the cell membrane.

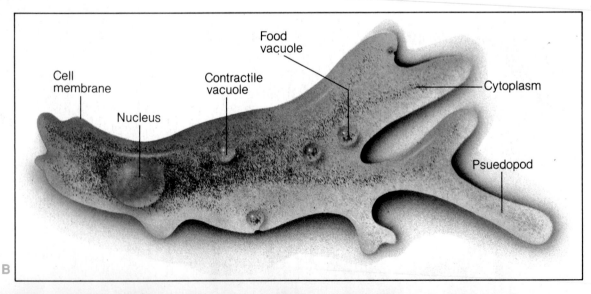

The process just described is used to transport "large" particles, such as bits of food or solid wastes, through the cell membrane. Other substances, such as water, oxygen, and carbon dioxide do not need to be carried in and out of the cell this way. These substances simply pass right through the cell membrane.

Protists that live in fresh water, such as many kinds of amebas, must deal with a tricky problem. Water tends to come into the cell from the environment. What do you think might happen to an ameba if excess water was allowed to build up in its cell? Fortunately, protists have a special cell structure that enables them to keep the right amount of water in the cell. This structure is called the contractile vacuole. Excess water collects in the saclike contractile vacuole. When the contractile vacuole is full, it contracts (hence its name) to squirt the water out of the cell.

Amebas reproduce by dividing into two new cells. Because amebas have a nucleus, the process of cell division involves more than simply making a copy of the hereditary material and then splitting in two. (This, you should recall, is the way in which monerans reproduce.) In amebas and all other cells with a nucleus, cell division involves a complex series of events.

Amebas respond in relatively simple ways to changes in their environments. They are sensitive to bright light and move to areas of dim light. Amebas are also sensitive to certain chemicals, moving away from some and toward others. How do you think these behaviors help amebas survive?

Figure 3–7 *Amebas, like most other protists, reproduce by dividing into two new cells.*

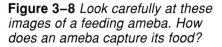

Putting the Squeeze On, p.174

Figure 3–8 *Look carefully at these images of a feeding ameba. How does an ameba capture its food?*

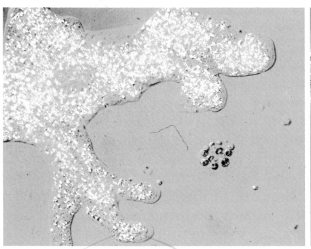

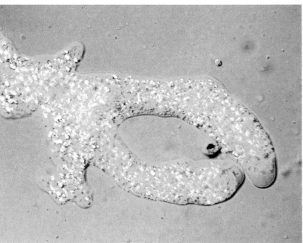

Ciliates

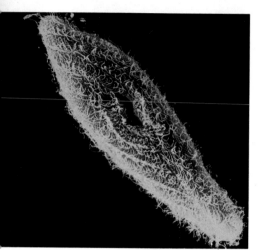

Ciliates have small hairlike projections called cilia (SIHL-ee-uh; singular: cilium) **on the outside of their cells.** The **cilia** act like tiny oars and help these organisms move. The beating of the cilia also helps to sweep food toward the ciliates. In addition, the cilia function as sensors. When the cilia are touched, the ciliate receives information about its environment.

Cilia may cover the entire surface of a ciliate or may be concentrated in certain areas. In some ciliates, the cilia may be fused together to form structures that look like paddles or the tips of paint brushes. A few ciliates possess cilia only when they are young. As adults, they attach to a surface and lose their cilia.

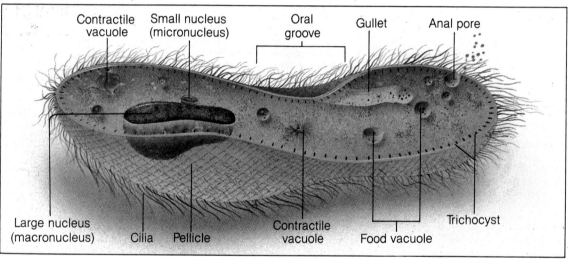

Contractile vacuole · Small nucleus (micronucleus) · Oral groove · Gullet · Anal pore

Large nucleus (macronucleus) · Cilia · Pellicle · Contractile vacuole · Food vacuole · Trichocyst

Figure 3–9 *Slipper-shaped* Paramecium *is probably the most familiar ciliate. What structures does a paramecium use in obtaining and digesting food?*

One of the most interesting ciliates is the **paramecium** (par-uh-MEE-see-uhm; plural: paramecia). As you can see in Figure 3–9, a paramecium has a tough outer covering called the pellicle (PEHL-ih-kuhl). The pellicle, which consists of the cell membrane and certain underlying structures, gives the paramecium its slipper shape. The cilia of the paramecium sweep food particles floating in the water into an indentation, or notch, on the side of the paramecium. This indentation, which is called the oral groove, leads to a funnellike structure known as the gullet. At the base of the gullet, food vacuoles form around the incoming food. Then the food vacuoles pinch off into the cytoplasm. As a food vacuole travels through the cytoplasm, the food is digested and the nutrients distributed. When the work of the food vacuole is completed, it joins with an area of the pellicle called the anal pore. The anal pore empties waste materials into the surrounding water.

Paramecia, like other ciliates, have two kinds of nuclei. The large nucleus controls the life functions of the cell. The small nucleus is involved in a complicated process called conjugation. During conjugation, two ciliates temporarily join together and exchange part of their hereditary material.

As you can see in Figure 3–11, paramecia reproduce by splitting in half crosswise. During this process, many cell structures are copied, divided, broken apart, and formed anew. Now take a moment to refer to the diagram of the paramecium in Figure 3–9 on page 66. Imagine that this paramecium is about to divide. Its left half is going to become one new cell, and its right half is going to become another new cell. What structures does the left half require in order to become a fully functional paramecium? What structures does the right half require?

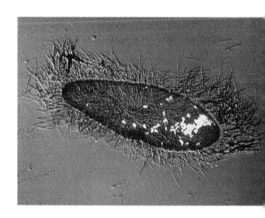

Figure 3–10 *When disturbed, a paramecium may release sharp spines known as trichocysts. How do these structures help the paramecium?*

O'DONNELL MIDDLE SCHOOL
211 CUSHING STREET
STOUGHTON, MA 02072

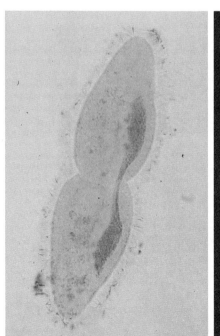
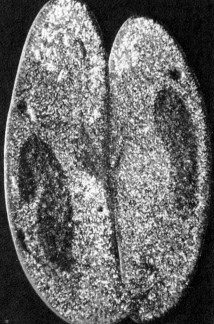

Figure 3–11 *Paramecia reproduce by dividing in half (left). This increases the number of paramecia. The process of conjugation also produces new paramecia but does not increase the number. The genetic material exchanged during conjugation may help to make one or both of the paramecia better able to deal with a changing environment (right).*

ACTIVITY

DOING

Capturing Food

1. Place one drop of a paramecium culture on a microscope slide. Use a depression slide if you have one.

2. Add one drop of *Chlorella*, a green algae, to the slide.

3. Locate a paramecium under the microscope using low power. Then switch to high power. What happened to the *Chlorella*?

4. "Feed" a tiny amount of carmine red granules or India ink to the paramecia. What happens to these inedible (impossible to eat) substances? What do you think will eventually happen to them?

How Big Is Big?

Protists range greatly in size. Some tiny flagellates are only 1 or 2 micrometers long. (A micrometer is one-millionth of a meter.) A paramecium is about 300 micrometers long. The fossil foraminiferan *Camerina* had a shell 10 centimeters wide. (A centimeter is one-hundredth of a meter.) Huge slime molds can be as large as 1 meter across.

Using a metric ruler and a calculator when necessary, find the numbers that make the following statements true:

1. I am _____ meters tall, or _____ centimeters tall, or _____ micrometers tall.

2. _____ zooflagellates 2 micrometers long placed end to end would equal 1 meter in length.

3. _____ *Camerina* shells placed end to end would equal 1 meter in length.

4. About _____ paramecia placed end to end could be placed across one *Camerina* shell.

5. _____ copies of this textbook laid end to end would equal 1 meter in length.

Zooflagellates

Flagellates (FLAJ-eh-layts) are protists that move by means of flagella. Recall from Chapter 2 that a **flagellum** is a long whiplike structure that propels a cell through its environment. Flagellates may be animallike, plantlike, or funguslike. Plantlike and funguslike flagellates will be discussed later.

Animallike flagellates are called zooflagellates. (The prefix *zoo-* means animal.) **Zooflagellates usually have one to eight flagella, depending on the species.** However, some types of zooflagellates have thousands of flagella.

Many zooflagellates live inside the bodies of animals. Can you explain why such zooflagellates are said to be symbiotic? Some symbiotic zooflagellates do not have any effect on their host. Others benefit their host and may even be essential to its survival. For example, termites and wood roaches rely on zooflagellates in their intestines to digest the wood that the insects eat. Without the zooflagellates, these wood-eating insects would quickly starve. Still other zooflagellates harm their host. Do you remember what this type of symbiotic relationship is called?

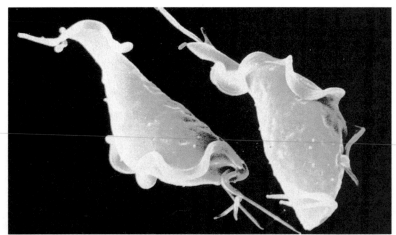

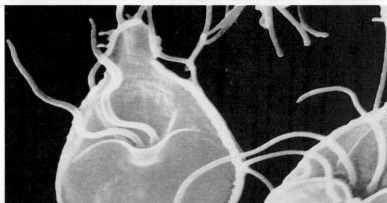

Figure 3–12 *Some zooflagellates cause diseases in humans.* Trichomonas *causes an infection of the female reproductive system (top).* Giardia *causes problems with the digestive system when it attaches to the walls of the small intestine (bottom).*

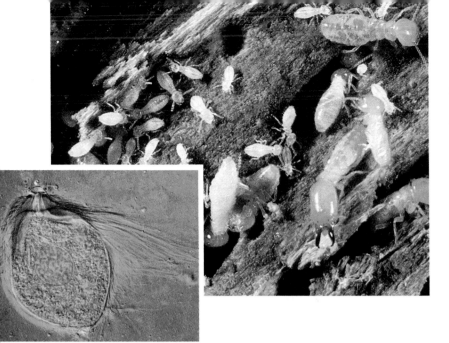

Figure 3–13 *Termites digest the wood they eat with the help of zooflagellates such as* Trichonympha *(inset). Why is the relationship between termites and* Trichonympha *an example of symbiosis?*

Parasitic zooflagellates are responsible for a number of diseases in humans and animals. One parasitic zooflagellate causes African sleeping sickness, which is transmitted from one host to another through the bite of the tsetse fly. Other kinds cause various types of diseases of the intestines.

Sporozoans

All sporozoans are parasites that feed on the cells and body fluids of their host animals. Many sporozoans have complex life cycles that involve more than one kind of host animal. During these life cycles, sporozoans form cells known as spores. The spores enable sporozoans to pass from one host to another. How does this happen? A new host may become infected with the sporozoans if it eats food containing the spores. Or it may become infected if it is bitten by a tick, mosquito, or other animal that has spores in its body.

Perhaps the most famous type of sporozoan is the organism that causes the disease malaria. This organism is *Plasmodium* (plaz-MOH-dee-uhm). *Plasmodium* has two hosts: humans and the *Anopheles* mosquito. Both hosts are necessary for *Plasmodium* to complete its life cycle.

When a mosquito infected with *Plasmodium* bites a human, it injects its saliva. The saliva contains substances that keep blood flowing so that the mosquito

CAREERS

Laboratory Technician

People who perform microscopic tests in laboratories are called **laboratory technicians.** They prepare blood and other body fluid samples by adding stains or dyes to them. After this is done, they examine the slide for abnormal cell growth or the presence of a parasite. Then laboratory technicians prepare reports on their findings.

Laboratory technicians also work in veterinary hospitals and help in the detection of infectious diseases in pets and farm animals. Some may work in agriculture to help study the effects of microorganisms on farm crops.

If you are interested in this career, write to the National Association of Trade and Technical Schools, 2251 Wisconsin Avenue NW, Washington, DC 20007.

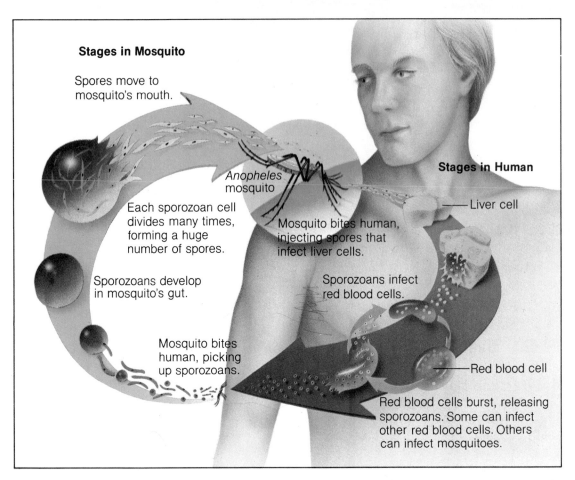

Stages in Mosquito

Spores move to
mosquito's mouth.

Anopheles
mosquito

Each sporozoan cell
divides many times,
forming a huge
number of spores.

Sporozoans develop
in mosquito's gut.

Mosquito bites
human, picking
up sporozoans.

Stages in Human

Mosquito bites human,
injecting spores that
infect liver cells.

Liver cell

Sporozoans infect
red blood cells.

Red blood cell

Red blood cells burst, releasing
sporozoans. Some can infect
other red blood cells. Others
can infect mosquitoes.

Figure 3–14 *The life cycle of*
Plasmodium *is quite complex. Why
is* Plasmodium *considered a
parasite? What are its hosts?*

can drink its fill. The saliva also contains *Plasmodium*
spores. The injected spores are carried by the bloodstream to the person's liver. There the spores divide
many times, producing a large number of sporozoan
cells. The resulting sporozoans are a form that can
infect red blood cells. After several days, these sporozoans burst out of the liver cells and invade red
blood cells, where they grow and multiply. Eventually
the sporozoans burst out of the red blood cells, destroying them.

Some of the sporozoans released when the red
blood cells burst can infect more red blood cells.
Others can infect mosquitoes. If the person with
malaria is bitten by another *Anopheles* mosquito, the
sporozoans along with the blood enter that
mosquito. In the mosquito, the sporozoans undergo
several stages of development and eventually form
spores, which continue *Plasmodium's* life cycle.

The malaria-causing sporozoans inside a patient's
body develop in such a way that all the millions of

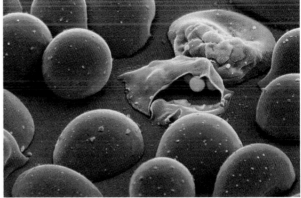

infected red blood cells burst at roughly the same time. Can you see why this is bad news for the patient? Within a few short hours, large amounts of parasites, bits of cells, and other kinds of "garbage" are dumped into the bloodstream. These released materials cause the patient with malaria to develop a high fever (about 41°C, or 105°F). As the patient's temperature climbs, the patient feels very cold and develops "goosebumps" all over. When the patient's temperature begins to return to normal, the patient feels very hot and sweats a great deal. And the disease has only just begun! The chills and fever of malaria occur again and again for weeks. In the most common kind of malaria, the chills and fever occur every 48 hours, last for about 6 to 8 hours each time, and persist for several weeks. Toward the end of that time, the chills and fever become less frequent and less severe. However, all may not be well even if the patient appears to have recovered. Several weeks after it first began, malaria often happens all over again.

Figure 3–15 *Humans contract the disease malaria through the bite of an infected mosquito (left). During the course of the disease, infected red blood cells burst at regular intervals, releasing* Plasmodium *cells that can infect other red blood cells or mosquitoes (right).*

3–2 Section Review

1. Briefly describe the four groups of animallike protists.
2. Compare the ways in which sarcodines, ciliates, and zooflagellates move.
3. Describe three ways in which animallike protists affect other organisms.

Critical Thinking—*Applying Concepts*
4. Explain why destroying the places where mosquitoes breed can help prevent malaria.

ACTIVITY
CALCULATING

Divide and Conquer

The ciliate *Glaucoma* reproduces by dividing into two cells. It reproduces at the fastest rate of all protists, dividing once every three hours under the best conditions. Assuming that conditions are perfect, how many times will *Glaucoma* divide during a day? If you start off with one cell that divides at the very start of the day, how many cells will you have at the end of the day if all of the cells survive?

CONNECTIONS

Revenge of the Protist

Bzzzzzzz! You hear a high-pitched whine, then feel an itch on your arm. You slap at the itch. Thwack! That mosquito is never going to bite anyone again.

Swatting is a great way to get rid of a few mosquitoes. But what do you do if you want to kill lots and lots of mosquitoes? No, bug spray is not a good answer.

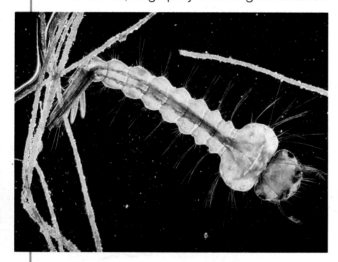

The poisons in bug spray *pollute* the environment, kill helpful insects such as bees and butterflies, and may be hazardous to your health. Although it may sound strange, a better answer is to use protists.

Scientists at the University of California at Berkeley recently discovered that a tiny ciliate called *Lambornella clarki (L. clarki)* can be a formidable foe to young mosquitoes. Young mosquitoes are wormlike, wingless, and legless creatures. They live at the surface of the water in quiet ponds, rain barrels, treeholes, and just about any other place that collects and holds water. The young mosquitoes feed on tiny particles that they filter from the water by using bristly mouthparts. These particles include small protists such as *L. clarki.*

To avoid being eaten, many microorganisms (microscopic organisms) change form. Some that are normally small and round make themselves large and flat. Others develop spines. *L. clarki* does something more amazing—it changes into a parasite that destroys the mosquitoes that feed on it.

Usually, *L. clarki* is a peaceful football-shaped ciliate that lives in treeholes and eats bacteria and other tiny bits of food. But when young mosquitoes are present, *L. clarki* cells become spherical, like a softball. These softball-shaped cells attach to the skin of young mosquitoes and then burrow into the body. The *L. clarki* cells multiply inside the body of their host, doing their deadly work. Eventually, the cells escape from the body of their dead or dying mosquito host. They can then infect other young mosquitoes. In nature, *L. clarki* cells can kill off all the young mosquitoes in a treehole. Having taken their revenge on the protist-devouring mosquitoes, the *L. clarki* cells resume life as peaceful football-shaped ciliates.

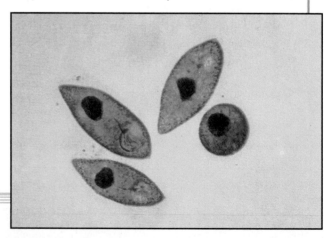

3-3 Plantlike Protists

Like other protists, plantlike protists are unicellular and most of them are capable of movement. **Like plants, plantlike protists are autotrophs that use light energy to make their own food from simple raw materials.** This food-making ability makes plantlike protists a vital part of the natural world. Can you see why? Many organisms rely directly on plantlike protists for food. Some of these organisms, such as animallike protists and tiny water animals, eat the plantlike protists. Others—certain animallike protists, sea anemones, corals, and giant clams, for example—are involved in symbioses with plantlike protists. The plantlike protists live inside their host's body and help to provide it with food. Still other animals rely indirectly on plantlike protists for food. For example, humans eat large fishes that eat smaller fishes that eat tiny animals that eat plantlike protists.

In addition to capturing energy and making it available to other organisms in the form of food, plantlike protists play another important role in the world. They produce oxygen as a byproduct of their food-making process. About 70 percent of the Earth's supply of oxygen is produced by plantlike protists.

Most kinds of plantlike protists are flagellates; that is, they move by means of flagella. To distinguish them from zooflagellates, plantlike flagellates are often called phytoflagellates. (The prefix

Guide for Reading

Focus on this question as you read.

▶ *What are plantlike protists?*

Figure 3-16 *Plantlike protists that do not belong to the three most important groups may still be rather interesting organisms. Some have elegant networks of glassy tubes that make up their skeleton (left). Others are covered with strange scales during their resting stage (top right and bottom right).*

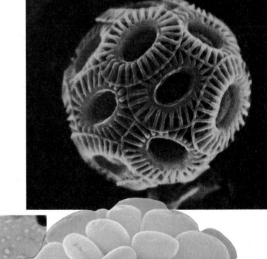

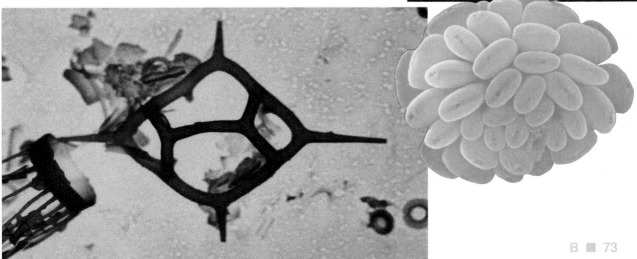

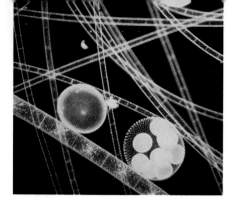

Figure 3–17 *Scientists do not agree where the dividing line between plantlike protists and plants should be placed. We classify long-stranded* Spirogyra *as a protist and spherical* Volvox *as a plant.*

Activity Bank

Shedding a Little Light on Euglena, p.176

phyto- means plant.) There are many different kinds of plantlike protists. In this section, you will read about three of the more interesting groups: **euglenas** (yoo-GLEE-nahz), **diatoms** (DIGH-ah-tahmz), and **dinoflagellates** (digh-noh-FLAJ-eh-layts).

Euglenas

Euglenas come in a variety of forms. Some are long and oval. Others are shaped like triangles, hearts, or tops. Still others live in branching colonies that look like bushes with oversized leaves. And a few live in cup-shaped "houses." Although euglenas are quite varied, most share three characteristics: a pouch that holds two flagella, a reddish eyespot, and a number of grass-green structures that are used in the food-making process. Scientists call the green food-making structures chloroplasts (KLOHR-oh-plasts).

One kind of euglena is shown in Figure 3–18. Like the paramecium you read about in the previous section, euglenas have a tough outer covering called the pellicle. The pouch on one end of the euglena holds two flagella, one long and one short. The long flagellum is used in movement. In the cytoplasm near the pouch is the reddish eyespot. The eyespot is sensitive to light. Why is it important for a euglena to be able to find light?

Figure 3–18 *The diagram shows the structure of a typical euglena. The pattern of grooves in a euglena's pellicle may be seen with the help of a scanning electron microscope (inset).*

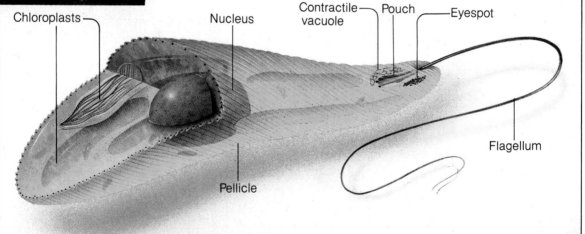

Chloroplasts — Nucleus — Contractile vacuole — Pouch — Eyespot — Pellicle — Flagellum

Diatoms

How would you feel about brushing your teeth with protists? Chances are you might not be too keen on the idea. But like it or not, you are probably doing exactly this every time you brush your teeth. Why? Because a part of many toothpastes is made from plantlike protists called diatoms.

Diatoms are among the most numerous of protists. There are about 10,000 living species of these aquatic (water-dwelling) organisms. And as you can see in Figure 3–19, diatoms are among the most attractive of organisms. Each diatom is enclosed in a two-part glassy shell. The shell looks like a tiny glass box or petri dish, with one side fitting snugly into the other. The two parts of the shell are covered with beautiful patterns of tiny ridges, spines, and/or holes. Imagine how surprised people must have been when they first looked at diatoms through a microscope and discovered lacy designs like those of stained-glass windows on tiny grains of sand!

When diatoms die, their tough glassy shells remain. In time, the shells collect in layers and form deposits of diatomaceous (digh-ah-tuh-MAY-shuhs) earth. Diatomaceous earth is a coarse, powdery

ACTIVITY READING

The Universe in a Drop of Water

Protist-sized, their entire world is a puddle of fresh water. They talk to diatoms and paramecia. They are friends with the glow-in-the-dark dinoflagellate *Noctiluca*. For them, a tiny shrimp or crayfish is an incredibly huge monster. Unbelievably, they are humans!

Discover a strange world of microscopic humans in the science-fiction short story "Surface Tension," by James Blish.

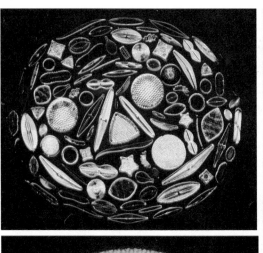

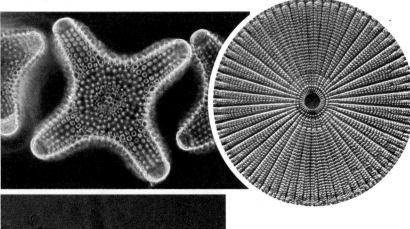

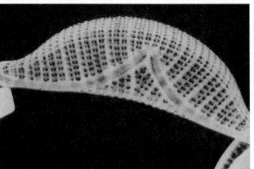

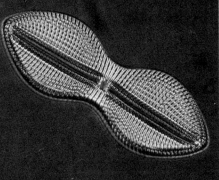

Figure 3–19 *The delicate glassy shells of diatoms are among the most beautiful forms in nature.*

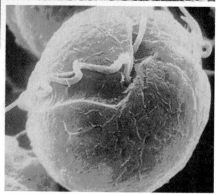

Figure 3–20 *Under certain conditions, dinoflagellates (inset) may form populations so huge that the water takes on their color. What are these population explosions called? What is their effect?*

material. This makes it an excellent polishing agent—and thus an important ingredient in toothpastes and car polish. Because diatomaceous earth reflects light, it is also added to the paint used to mark traffic lanes on roads and highways. Why do you think diatomaceous earth might be added to scouring cleansers?

Dinoflagellates

In August of 1987, the waters off the southwestern coast of Florida suddenly turned yellow and brown. Soon thousands of dead fish washed up on shore. Two months later, the same thing happened along much of the North Carolina coast. Tourism and fishing industries in that state were hard hit by the poisoned, discolored waters of what is called the red tide. A variety of illnesses in at least 41 people was another result. This was not the first time a red tide had struck the United States, or for that matter, other places in the world. Red tides have swept onto beaches in New England, Los Angeles, Taiwan, Guatemala, Korea, and Tasmania, to name a few.

Red tides occur when protists known as dinoflagellates reproduce so rapidly that the water becomes colored by them. Although they are called red tides, such occurrences are not necessarily red in color. This is because dinoflagellates range in color from yellow-green to orange-brown. Red tides can be dangerous because some dinoflagellates produce toxins (poisons) that can injure or even kill living things.

If you observed dinoflagellates through a microscope, you would notice that most have cell walls that look like plates of armor. They also have two flagella that propel them through the water. One flagellum trails from one end like a tail. The other wraps around the middle of the organism like a belt. The movement of this flagellum causes the dinoflagellate to spin like a tiny top.

Some dinoflagellates have a characteristic that has amazed sailors since they first set sail on the world's oceans. On some nights, ocean waves glitter with thousands of tiny sparkles or glow with an eerie bluish light. The twinkling or glowing lights are produced by dinoflagellates. This glow is similar to the glow produced by fireflies, or lightning bugs.

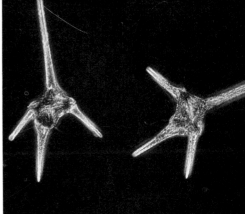

3–3 Section Review

1. Name and briefly describe three groups of plantlike protists.
2. Discuss three ways in which plantlike protists affect other living things.
3. How does a euglena detect light?
4. What is a red tide?
5. Why do experts disagree about how euglenas should be classified?

Connection—*Language Arts*

6. The Greek word *dinos* means to whirl or rotate. Why is dinoflagellate an appropriate name?

Figure 3–21 *A few dinoflagellates, such as* Noctiluca *(top right), glow at night. Most dinoflagellates have elaborate armored cell walls (top left).* Ceratium *(bottom) gets its name from a Greek word meaning horn or antler. Can you see why the name is appropriate?*

3–4 Funguslike Protists

Funguslike protists can have a great effect on human life, as the Great Potato Famine in Ireland illustrates. In 1845 and 1846, one type of funguslike protist destroyed the entire potato crop, creating a famine that caused the deaths of about one third of Ireland's population. Funguslike protists can also attack other crops, such as cabbages, corn, and grapes. Animals get diseases caused by funguslike protists. The fuzzy white growths that sometimes appear on the fins and mouth of aquarium fishes is one such disease.

Guide for Reading

Focus on this question as you read.

▶ How do slime molds differ from other protists?

Figure 3–22 *The fuzzy white growth that covers this dead aquarium fish is the funguslike protist that killed it.*

Although it is clear that funguslike protists are important, it is not clear exactly what they are. But we do know about their characteristics. Funguslike protists are heterotrophs. Most have cell walls, although a few lack a cell wall altogether. Some are almost identical to amebas during certain stages in their life cycles. Many have flagella at some point in their lives. Because of characteristics like these, many experts classify these puzzling organisms as protists.

One of the more interesting types of funguslike protists are the **slime molds.** At one point in their life cycle, these protists are moist, flat, shapeless blobs that ooze slowly over dead trees, piles of fallen leaves, and compost heaps. A few slime molds in this stage of life are shown in Figure 3–23. Can you see why the name slime molds is appropriate for these organisms?

Reproduction in slime molds involves the production of a structure called a fruiting body, which contains spores. Spores are special cells that are encased in a tough protective coating. Each spore can develop into a new organism.

Figure 3–23 *Unlike most protists, slime molds are visible to the unaided eye at some stages of their life cycle. In one phylum of slime molds, a single amebalike cell may grow into an enormous cell that contains many nuclei (left). In the other phylum, many amebalike cells come together to produce a large mass of cells that behaves like a primitive multicellular organism (right).*

Spores released by a slime mold develop into small amebalike cells. In one phylum of slime molds, each amebalike cell may develop into a single huge cell several centimeters in diameter. This large cell contains many nuclei. Eventually, the large cell settles in one place and produces fruiting bodies.

In the other phylum of slime molds, the small amebalike cells live independently for a while and reproduce rapidly. When the food supply is used up, groups of the amebalike cells gather together to produce a large mass of cells. This mass of cells begins to function as a single organism. Could this have

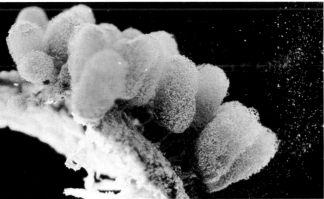

Figure 3–24 *The reproductive structures in slime molds, which are known as fruiting bodies, contain spores. What do the spores become?*

been a way in which many-celled organisms evolved from single-celled ones? Some scientists believe it could. This unusual behavior forces scientists to stretch the definition of protists. Protists are defined as being unicellular—but here is a group of protists acting like a primitive multicellular (many-celled) organism!

The solid mass of cells may travel for several centimeters. It then forms a fruiting body that produces spores. These spores develop into amebalike cells that continue the cycle.

The slime molds that form multicellular structures are interesting to biologists who study how cells communicate. The formation of a complex structure like the fruiting body from what was formerly a group of independent cells is an intriguing process. It has kept biologists busy for decades, and its secrets are still not fully understood.

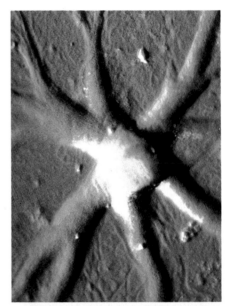

3–4 Section Review

1. How do slime molds differ from other protists?
2. What are the characteristics of funguslike protists?
3. Draw a flowchart that shows the life cycles of the two types of slime molds.
4. How do funguslike protists affect humans?

Critical Thinking—*Expressing an Opinion*

5. Should slime molds be placed in a kingdom by themselves? Why? What sort of information would you like to have in order to make a better, more informed decision on this matter?

Figure 3–25 *In one phylum of slime molds, small amebalike cells come together (top) to form a mass of cells that behave like a single organism (bottom).*

Laboratory Investigation

Examining a Slime Mold

Problem

What are the characteristics of a slime mold?

Materials *(per group)*

large glass beaker	crushed oatmeal flakes
small glass bowl	
paper towel	medicine dropper
filter paper contain-ing slime mold	magnifying glass
	dissecting needle

Procedure 🧪 ⬛

1. Wrap the small glass bowl with a paper towel so that the mouth of the bowl is covered by a smooth flat paper surface.

2. Place the covered bowl in the beaker so that the mouth of the bowl faces up.

3. Partially fill the beaker with water so that the water level is about three fourths of the way up the sides of the bowl.

4. Place the small piece of filter paper containing the slime mold in the center of the paper towel that covers the bowl.

5. Sprinkle a tiny amount of crushed oatmeal flakes next to the piece of filter paper.

6. Using the medicine dropper, add two to three drops of water to the slime mold and oatmeal flakes. Set in a cool, dark place.

7. Examine the beaker each day for three days. Record your observations.

8. After three days, remove the glass bowl from the beaker. Place the bowl on your work surface.

9. Using your magnifying glass, examine the slime mold.

10. With a dissecting needle, puncture a branch of the slime mold. **CAUTION:** *Be careful when using a dissecting needle.* Observe the slime mold for a few minutes.

Observations

1. Describe the changes that took place in the slime mold during the three-day observation period.

2. What activity did you observe in the slime mold when you examined it with the magnifying glass?

3. Describe what happened to the puncture that you made in the slime mold.

Analysis and Conclusions

1. Explain why oatmeal was sprinkled on the paper towel.

2. Is the slime mold a heterotroph or an autotroph? Explain.

3. Based on your observations, describe the characteristics of a slime mold.

4. **On Your Own** Design an experiment to determine the response of the slime mold to substances such as salt or sugar.

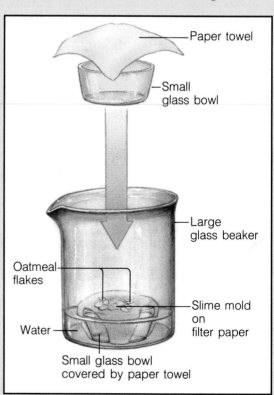

Paper towel

Small glass bowl

Large glass beaker

Oatmeal flakes

Slime mold on filter paper

Water

Small glass bowl covered by paper towel

Summarizing Key Concepts

3–1 Characteristics of Protists

▲ Protists are microscopic, unicellular organisms that have a nucleus and a number of other specialized cell structures.

▲ According to one hypothesis, the first protists were the result of a symbiosis among several types of bacteria.

▲ Although there is much debate about their proper classification, protists can be grouped in three general categories. These are: animallike protists, plantlike protists, and funguslike protists.

3–2 Animallike Protists

▲ Like animals, animallike protists are heterotrophs that can move and that are made up of cells that contain a nucleus and lack a cell wall.

▲ Animallike protists are divided into four main groups: sarcodines, ciliates, zooflagellates, and sporozoans.

▲ Sarcodines have pseudopods.

▲ Ciliates are characterized by cilia.

▲ Flagellates move by means of flagella.

▲ Sporozoans are parasites.

3–3 Plantlike Protists

▲ Plantlike protists are autotrophs that use light energy to make their own food from simple raw materials.

▲ Many organisms rely on plantlike protists for food.

▲ About 70 percent of the Earth's supply of oxygen is produced by plantlike protists.

3–4 Funguslike Protists

▲ Some funguslike protists cause diseases in crops or in animals.

▲ Funguslike protists are heterotrophs.

Reviewing Key Terms

Define each term in a complete sentence.

3–1 Characteristics of Protists
protist

3–2 Animallike Protists
sarcodine
ciliate
zooflagellate
sporozoan
pseudopod
ameba
cilia
paramecium
flagellum

3–3 Plantlike Protists
euglena
diatom
dinoflagellate

3–4 Funguslike Protists
slime mold

Chapter Review

Content Review

Multiple Choice

Choose the letter of the answer that best completes each statement.

1. Which of the following is characteristic of most protists?
   a. They are unable to move on their own.
   b. They can be seen with the unaided eye.
   c. They lack a nucleus and many other cell structures.
   d. They are unicellular.
2. Which structure helps a freshwater protist get rid of excess water?
   a. food vacuole   c. macronucleus
   b. contractile vacuole   d. cilium
3. Malaria is caused by a type of
   a. sporozoan.   c. dinoflagellate.
   b. sarcodine.   d. zooflagellate.
4. Which of the following uses cilia to move?
   a. euglena   c. paramecium
   b. ameba   d. *Plasmodium*
5. Which of the following is considered to be an animallike protist?
   a. slime mold   c. euglena
   b. ameba   d. dinoflagellate

6. Animallike protists
   a. have a thick cell wall.
   b. produce about 70 percent of the Earth's supply of oxygen.
   c. are responsible for red tides.
   d. are heterotrophs.
7. Which is not a type of plantlike protist?
   a. zooflagellate   c. phytoflagellate
   b. dinoflagellate   d. diatom
8. Radiolarians, foraminiferans, and amebas belong to the group of protists known as
   a. ciliates.   c. sporozoans.
   b. sarcodines.   d. euglenas.
9. Slime molds
   a. are always unicellular.
   b. are autotrophs.
   c. are considered to be animallike protists.
   d. form structures called fruiting bodies.

True or False

If the statement is true, write "true." If it is false, change the underlined word or words to make the statement true.

1. <u>Funguslike</u> protists are also known as protozoa.
2. Each euglena contains one or more grass-green <u>macronuclei</u>.
3. A radiolarian captures food by using its <u>pseudopods</u>.
4. African sleeping sickness is caused by <u>ciliates</u>.
5. The first protists probably developed from a <u>symbiosis</u> among several kinds of bacteria.
6. All sporozoans are <u>producers</u>.
7. Paramecia swim using <u>flagella</u>.

Concept Mapping

Complete the following concept map for Section 3–1. Refer to pages B6–B7 to construct a concept map for the entire chapter.

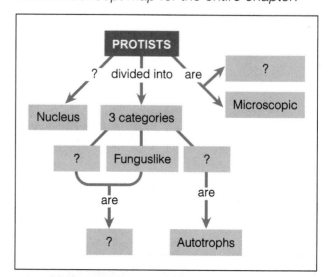

Concept Mastery

Discuss each of the following in a brief paragraph.

1. Name the group of protists that is most closely linked with each of the following:
   a. chalk
   b. diatomaceous earth
   c. red tides
   d. malaria
   e. African sleeping sickness
   f. turning chloroplasts on and off
   g. two kinds of nuclei
   h. fruiting bodies
   i. pseudopods
   j. two-part glassy shell
   k. potato famine
   l. digesting wood
2. How are sarcodines similar to one another? How are they different?
3. Explain how a paramecium catches the bacteria on which it feeds. Then describe the cell structures a captured bacterium would encounter from the time it enters the paramecium cell to the time the digested bacterium is expelled from the cell.
4. What protist causes malaria? How is this protist transmitted from one host to another?
5. Tiny plantlike protists are found in the cells of certain radiolarians. How do these plantlike protists help their host? How might the radiolarian help its "guests"?

Critical Thinking and Problem Solving

Use the skills you have developed in this chapter to answer each of the following.

1. **Relating cause and effect** Certain kinds of chemical wastes cause plants and plantlike organisms to grow at an extremely rapid rate. With this in mind, how might pollution cause red tides?
2. **Applying concepts** Examine the protist in the accompanying photograph. In what group of protists should this organism be placed? Explain. What sort of information would help you to classify this protist?
3. **Making predictions** What would happen if all the plantlike protists were to vanish from the face of the Earth? Explain.
4. **Evaluating theories** When a certain amebalike protist is treated with a bacteria-killing antibiotic, something strange happens. The small rod-shaped structures located around each of its many nuclei disappear, and the protist soon dies. Does this finding support the hypothesis that protists evolved from a symbiosis among several kinds of bacteria? Why or why not? What additional information would you like to have in order to be more sure of your answer?
5. **Using the writing process** Imagine that you are the protist of your choice. Write a letter to a potential pen pal describing yourself, where you live, your hopes and dreams, and whatever else you think is important.

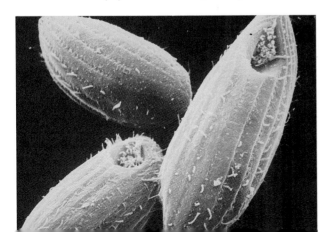

Fungi

Guide for Reading

After you read the following sections, you will be able to

4-1 Characteristics of Fungi
- Describe the major characteristics of fungi.

4-2 Forms of Fungi
- Compare mushrooms, yeasts, and molds.

4-3 How Fungi Affect Other Organisms
- Discuss ways in which fungi interact with other living things.

Early on a spring morning, a large pampered pig is pushed in its own wheelbarrow into an oak forest near Perigord, France. There its master gently puts the pig on a leash. The pig is now ready for the hunt!

Soon the pig catches a whiff of a wonderful odor—one that is too faint for people to smell. The pig begins to dig, but its master quickly stops it. He does not want the animal to destroy the buried treasure it has found. This "treasure" is not gold or silver. It is an ugly round black fungus known as a truffle. Because of its delicious flavor, this thick-skinned, warty cousin of the mushroom is considered a delicacy. In fact, truffles can sell for more than $1400 a kilogram!

Truffles are just one of the many different kinds of fungi. What are fungi? What do they look like? How do they affect humans and other living things? Read on to discover more about the strange world of the kingdom Fungi.

Journal *Activity*

You and Your World   Have you ever eaten a mushroom? What do you remember about the first time you ate a mushroom? Has your opinion of mushrooms changed since your first tasting? In your journal, explore your thoughts and feelings about this edible fungus.

The sensitive nose of a trained pig can detect truffles buried beneath the soil. Dug up, scrubbed, and cooked, truffles become expensive taste treats.

4–1 Characteristics of Fungi

Unnoticed, a speck of dust lands on the back of an ant. But this is no ordinary dust—it is alive! Tiny glistening threads emerge from the dust and begin to grow into the ant's body. As they grow, the threads slowly devour the ant while it is still alive. Chemicals released by the threads dissolve the ant's tissues. The threads absorb the dissolved tissues and use the energy from them to spread further into the ant's body. Within a few days, the ant's body is little more than a hollow shell filled with a tangle of the deadly threads. Then the threads begin to grow up and out of the dead ant, winding together to produce a long stalk with a knobby lump at its tip. The stalked structure releases thousands of dustlike spores, which are carried by the wind to new victims.

The strange ant killer is a **fungus** (FUHN-guhs; plural: fungi, FUHN-jigh). Fungi range in size from tiny unicellular (one-celled) yeasts to huge tree fungi over 140 centimeters long. Fungi may look like wisps of gray cotton, white volleyballs, tiny brightly colored umbrellas, blobs of melted wax, stubby fingers of yellow-green slime, or miniature red bowls. (And there are many that defy description!) Although fungi come in a variety of shapes, sizes, and colors, they share many important characteristics. They are similar in the way they get their food, in their structure, and in the way they reproduce.

Figure 4–1 *When the insect-killing fungus has completed its deadly work, it produces stalked fruiting bodies that grow from the empty husks of its insect victims. What is the function of the fruiting bodies?*

"Feeding" in Fungi

All fungi are heterotrophs (organisms that cannot make their own food). They obtain the energy and chemicals they need by growing on a source of food. **Fungi release chemicals that digest the substance on which they are growing and then they absorb the digested food.** (Animals, on the other hand, first take in—eat—their food and then digest it.)

Some fungi capture small animals for food. Oyster mushrooms, for example, release a chemical that stuns tiny roundworms in the soil. Certain threadlike soil fungi have tiny nooses that they use to snare their roundworm prey. Once a roundworm has been captured, the fungus begins to grow on it. Can you describe what happens after the fungus starts to grow on its prey?

Other fungi obtain their food through symbiotic relationships. (Remember that a close relationship between two kinds of organisms in which at least one of the organisms benefits is known as a symbiosis.) Some symbiotic fungi, like the ant-killing fungus you just read about, are parasites that harm their host. Other symbiotic fungi help their host. Later in this chapter, you will read about some specific examples of harmful and helpful symbiotic fungi.

Many species of fungi get their food from the remains of dead organisms. These fungi are decomposers. Recall from Chapter 2 that decomposers break down dead plant and animal matter. The broken-down products become the foods of other living things. Why can such fungi, along with certain bacteria, be called "the Earth's cleanup crew"?

Figure 4–2 *The scarlet cup fungi, purple coral fungi, and whitish earthstar fungus display just a few of the many colors and shapes of fungi.*

Figure 4–3 *An unlucky roundworm struggles in vain as the nooses of a soil fungus tighten around it in this microscopic drama.*

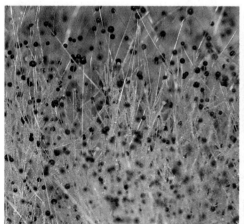

ACTIVITY

Making Models of Multicellular Organisms

You can get a better idea of how different the structure of fungi is from that of other multicellular organisms by constructing a model. For these models you will need some sugar cubes and thin licorice whips.

1. For your model of a fungus, tangle and twist the licorice whips together to form a compact structure.

2. For your model of a plant or animal, stack the sugar cubes to produce a compact structure that is several cubes in length, height, and depth.

What do the licorice whips represent? What do the sugar cubes represent?

■ How is the basic structure of a fungus different from that of other multicellular organisms?

Structure of Fungi

A few fungi, such as the yeast used by bakers, are unicellular (one-celled). Most fungi, however, are multicellular (many-celled). The fat white toadstools at the base of a tree, the black spots of mildew on a shower curtain, and the fuzzy greenish mold on a piece of old cheese are all examples of multicellular fungi.

Multicellular fungi are made up of threadlike tubes called hyphae (HIGH-fee; singular: hypha). The **hyphae** branch and weave together in various ways to produce many different shapes of fungi. Hyphae can grow quite quickly—certain fungi can produce about 40 meters of hyphae in an hour. This is part of the reason mushrooms seem to pop up in fields and lawns overnight!

Interestingly, fungi are not multicellular in the same way that plants and animals are. All of the cells that make up the bodies of plants and animals are distinct units. Each plant or animal cell contains one nucleus (although there are some exceptions) and is enclosed by a cell membrane that separates it from other cells. The hyphae of fungi, on the other hand, are continuous threads of cytoplasm that contain many nuclei. Now can you explain why substances move more quickly and freely in fungi than in other multicellular organisms?

Figure 4–4 Fungi are made up of threadlike structures called hyphae (right). A close examination of a bread mold reveals that it is made of white, threadlike hyphae peppered with round black spore cases (left). The thick, heavy plates of shelf fungi are also made up of hyphae (center). In which kind of fungi are the hyphae more closely packed together?

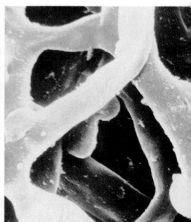

The hyphae of some fungi are divided into compartments by incomplete cross walls. Although these compartments are traditionally referred to as "cells," they are not enclosed by a cell membrane. Nuclei and other cell structures move quite freely through the openings in the cross walls. The only "real" cells are reproductive cells located at the tips of some of the hyphae.

Reproduction in Fungi

Many fungi reproduce by means of spores. Fungal **spores** are tiny reproductive cells that are enclosed in a protective cell wall. Because they are very small and light in weight, spores can be carried great distances by the wind. If a spore lands in a place where growing conditions are right, it can sprout and develop hyphae. How does this fact help to explain why fungi are found just about every place in the world?

Fungi produce spores in a special structure called the fruiting body. Some fungi have simple fruiting bodies that consist of a stalk with a cluster of spores at its tip. Other fungi—mushrooms, cup fungi, and puffballs, for example—have large, complex fruiting bodies made up of many closely packed hyphae. A

Activity Bank
Spreading Spores, p. 177

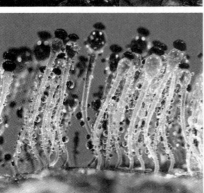

Figure 4–5 *Puffballs release a cloud of spores when touched (right). The girl posing beside the giant puffball (above) would be well advised to avoid touching the fungus! The spores of puffballs are spread by the wind. Some fungi spread their spores in other ways. The lacy stinkhorn produces a fluid that smells like rotting meat (top left). When a fly eats the fluid, it also takes in spores. The spores pass unharmed through the fly's body and are deposited over great distances. Tiny* Pilobolus *can throw its spore cases as much as a meter away (bottom left)! This is roughly equivalent to throwing a baseball the length of several football fields.*

single fruiting body such as a large puffball may produce trillions of spores.

So why aren't we surrounded by millions of mushrooms or buried in puffballs? The answer is simple: Very few fungal spores find the proper combination of temperature, moisture, and food that they need to survive. Even fewer young fungi survive long enough to produce spores of their own. Can you now explain why fungi produce huge numbers of spores?

4–1 Section Review

1. Briefly discuss the basic characteristics of fungi.
2. What are hyphae?
3. How do fungi obtain food?
4. Describe how fungi reproduce.

Critical Thinking—*Relating Concepts*
5. Explain why it is important for fungi (and other organisms that cannot move) to produce offspring that can easily travel from one place to another.

Guide for Reading

Focus on this question as you read.

▶ *How are mushrooms, yeasts, and molds similar? How are they different?*

4–2 Forms of Fungi

Fungi, like other organisms, are classified in a way that best shows the evolutionary relationships among the members of the group. As you might expect for such a strange and diverse group of organisms, the guidelines for classifying fungi are rather complex. As a result, people find it useful to group fungi according to their basic form, or shape. It is important to note that these groupings are not the formal classifications you learned about in Chapter 1. There are three basic forms of fungi: mushrooms, yeasts, and molds. **Mushrooms are shaped like umbrellas. Yeasts consist of single cells. And molds are fuzzy, shapeless, fairly flat fungi that grow on the surface of an object.**

Figure 4–6 *Although a morel looks, smells, and tastes like a mushroom, it is more closely related to baker's yeast than to mushrooms.*

Although these three categories are handy for everyday purposes, they do not make up a perfect organizational system. Why? Some fungi have very unusual shapes and cannot be placed in any of these categories. Others have more than one shape. The fungus that causes the disease known as thrush, for example, occurs both as a yeast and as a mold. Still others appear to have the correct shape for a group but are not placed in that group. Even though the morel shown in Figure 4–6 looks (and tastes) a lot like a true mushroom, it is not a mushroom.

Mushrooms

Have you ever ordered a pizza with all the trimmings? If so you ate fungi known as **mushrooms.** Figure 4–7 shows some mushrooms that might be encountered on a walk through the woods. As you can see, a mushroom has a stemlike structure called a stalk. In many types of mushrooms, the stalk is decorated with a structure called a ring, which looks somewhat like a very short skirt. On top of the stalk is the mushroom's cap. The mushroom's spores are produced on the underside of the cap. The spores are often located on thin sheets of tissue called gills, which extend from the stalk to the outer edge of the cap.

Activity Bank

Yeast Meets Best, p. 178

Figure 4–7 *Most familiar mushrooms produce their spores on gills. Some mushrooms, however, produce their spores in tubes and release them through pores. Others bear their spores on tiny flaps known as teeth.*

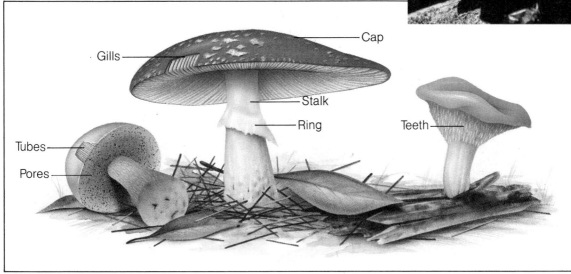

PROBLEM Solving

A Hot Time for Yeast

Yeasts are tiny unicellular fungi. Some yeasts are used to make alcoholic beverages and foods such as bread. The accompanying graph shows the effect of temperature on the level of activity of yeast cells. Use the graph to answer the questions that follow:

Interpreting Graphs

1. How does temperature affect yeast activity?

2. At what temperature is the yeast activity the highest?

3. At what temperature does the yeast activity decrease sharply?

4. Why is yeast dissolved in warm water when it is used to make bread?

5. Explain why bread dough is sometimes placed in a slightly warm oven for an hour or so before it is made into a loaf and baked in a hot oven.

6. Would you expect to find live yeast cells in a slice of bread? Explain.

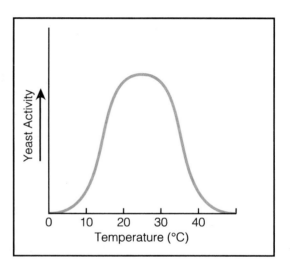

Figure 4–8 *Yeasts are used to produce bread, fuel, vitamins, chemicals, and even medicines such as the vaccine for hepatitis B. Yeasts reproduce by budding. A round scar results when a bud breaks off from its parent cell. How many buds has the larger yeast cell produced?*

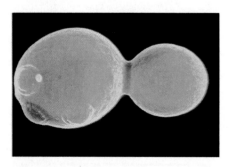

Yeasts

Most people cannot help but stop and take a deep breath when they pass a bakery. There is something about the smell of fresh bread that excites the senses. The next time you pass a bakery you might whisper a soft thank-you to another type of fungi, the **yeasts.**

In order to make soft, fluffy bread, bakers add yeast to the flour, water, sugar, salt, and other ingredients that make up bread dough. The bakers then allow the dough to sit for a while in a warm place. Bread dough is a great environment for yeast— moist, warm, and full of food. As it grows, the yeast produces carbon dioxide gas. The carbon dioxide gas forms millions of tiny bubbles in the dough. You see these bubbles as holes in a slice of bread. What do you think would happen if a baker forgot to put yeast in the bread dough?

Unlike other fungi, yeasts may reproduce by a process known as budding. During budding, a portion of the yeast cell pushes out of the cell wall and forms a tiny bud. In time, the bud breaks away from the parent cell and becomes a new yeast.

Molds

Centuries ago people sometimes treated infections in a rather curious way. They placed decaying breads, cheeses, or fruits on the infection. Although the people did not have a scientific reason for doing this, every once in a while the infection was cured. What these people did not and could not know was that the cure was due to a type of **mold** that grows on certain foods.

In 1928, the Scottish scientist Sir Alexander Fleming found out why this treatment worked. Fleming discovered that a substance produced by the mold *Penicillium* could kill certain bacteria that caused infections. Fleming named the substance penicillin. Since that time penicillin, an antibiotic, has saved millions of lives.

Molds are used to make many foods, such as tofu (bean curd), soy sauce, and cheeses. The blue streaks in blue cheese, for example, are actually

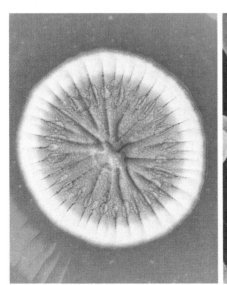
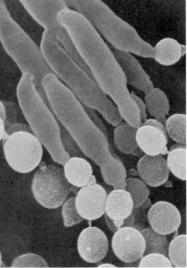

ACTIVITY
DOING

Making Spore Prints

1. Place a fresh mushroom cap, gill side down, on a sheet of white paper. Cover with a large glass jar, open end down. **CAUTION:** *If you use a wild mushroom, wash your hands thoroughly after handling it. Do not leave wild mushrooms where small children can reach them. Some wild mushrooms are poisonous.*

2. After several days, carefully remove the jar and lift off the mushroom cap. You should find a spore print on the paper.

3. Very carefully spray the paper with clear varnish or hair spray to make a permanent spore print.

4. Examine the print with a magnifying glass.

5. (Optional) Prepare a spore print using a different kind of mushroom.

What color are the spores in your spore prints? How are your spore prints similar? How are they different? How do your spore prints compare to ones prepared by your fellow classmates?

Figure 4–9 *The mold* Penicillium *(left) produces spores at the tips of tiny branches (right). What important antibiotic comes from* Penicillium*?*

mold. Of course, not all molds help to make foods or provide valuable medicines. Most molds are just plain, ordinary fungi. So if you discover fuzzy growths of mold on decaying breads, cheeses, or fruits, you should probably just clean out the refrigerator or take out the trash.

4–2 Section Review

1. How is a yeast different from a mushroom?
2. List five different uses for yeasts.
3. Discuss three ways in which molds affect people.

Critical Thinking—*Evaluating Classification Schemes*
4. Explain why fungi are not divided into phyla according to basic shape.

Guide for Reading

Focus on these questions as you read.

▶ How do fungi harm other organisms?

▶ How do fungi help other organisms?

ACTIVITY

READING

A Secret Invasion

What might mushrooms say if they could speak? Are these small umbrella-shaped organisms as innocent as they seem? For one poet's answers to these questions, read the poem "Mushrooms," by Sylvia Plath.

4–3 How Fungi Affect Other Organisms

Fungi interact with other organisms in many different ways. Some fungi harm other organisms. Such fungi are often disease-causing parasites of plants and animals. Other fungi are helpful to other organisms and may even be necessary for their survival.

Fungi and Disease

Have you ever looked closely at an apple and noticed a sprinkling of small, hard brown "scabs" on its skin? Or have you ever seen round black or gray spots on the leaves of a lilac or rose bush? These scabs and spots are the result of plant diseases caused by fungi. And although the apple is still safe to eat, the lilac or rose plant might be in serious trouble.

Scabs and spots are not the only signs of fungal diseases. Some disease-causing fungi make the stem, roots, or fruit of crop plants rot, or decay. Others, such as those that cause Dutch elm disease and

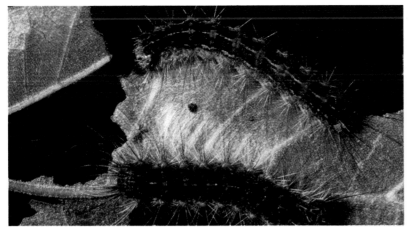

chestnut blight, kill trees that are prized for their beauty and wood. A few change kernels of growing grain into bags of useless fungal spores. Still others damage stored crops such as wheat, corn, oats, peanuts, and rice, making them unfit to eat. Can you now explain why farmers and gardeners spend millions of dollars on fungicides (fungus-killing substances) each year?

In addition to damaging or even completely destroying crops, some fungi that infect plants produce toxins (poisons) that can injure or even kill humans and animals. For example, one fungus that grows on stored grain produces a toxin that is one of the most powerful cancer-causing substances. (Scientists know this toxin as aflatoxin.) In small doses, this toxin can cause liver cancer. In large doses, it is fatal. Like most fungi, the toxin-producing fungus will grow only if there is sufficient moisture. Why is it important to thoroughly dry crops such as peanuts or corn before they are stored?

Figure 4–10 *Fungi affect other organisms in many ways. Leafcutter ants grow the fungi they eat on bits of plants they carry to their underground nests (top right). Gypsy moth caterpillars (left), which cause a great deal of damage to trees, can be controlled with the help of a parasitic fungus. Grains of corn are changed into bags of toxic spores by the fungus known as corn smut (bottom right).*

Another fungus replaces grains of rye with hard spiky poisonous growths known as ergot. People who eat bread or other grain products containing ergot may experience burning or prickling sensations, hallucinations, and convulsions. In extreme cases, the flow of blood to the arms and/or legs may be cut off, resulting in infections and possible loss of the affected limbs.

Ergot poisoning can be fatal—although not necessarily through the direct actions of the toxins. Some historians have suggested that the witchcraft trials in Salem, Massachusetts, may have been due in part to people's terror at the strange symptoms of ergot poisoning. Early in 1692, several girls in Salem

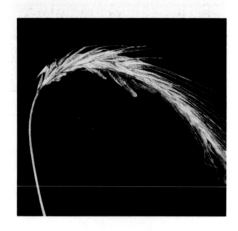

Figure 4–11 *The long, curved spikes on this head of rye are ergot. What is the connection between fungi and the Salem witchcraft trials?*

began suffering from pains and odd behaviors that were thought to be caused by witchcraft. By the end of the year, 19 people had been hanged as witches as a result of the ensuing witch hunt and trials.

Animals, like plants, can become infected by fungal diseases. Fungi cause a number of severe, sometimes fatal, lung diseases in poultry (chickens, ducks, and other kinds of farm birds). They can also produce itchy or painful sores on the skin of pets such as dogs, cats, and birds. While most fungal diseases of animals are troublesome, a few have proven to be useful to humans. Plant-eating pests such as gypsy moth caterpillars, mites, and aphids can be killed by deliberately infecting them with certain fungi. These fungal pesticides are a lot more effective than chemicals—and a lot safer for the environment, too!

Although some human fungal diseases are serious, most are simply annoying. One causes the fingernails or toenails to grow in crooked and/or fall out. The diseases known as ringworm and athlete's foot cause itchy, reddened, and raw patches on the skin.

Fungus-Root Associations

As you have discovered, plants are plagued by a host of fungal diseases. However, plants get help against some of these diseases from—you guessed

Figure 4–12 *Some mycorrhizae send up mushrooms or puffballs. This sometimes produces a "fairy ring" around a tree. Folklore has it that trees surrounded by fairy rings are the favorites of fairies, so chopping down such trees is extremely bad luck.*

it—other fungi. Helpful fungi coat the roots of about 80 percent of the world's plants. Some of these fungi simply cover the surface of the roots. Others actually send hyphae into the roots' cells. Scientists give these fungus-root associations a fancy name: mycorrhizae (migh-koh-RIGH-zee; singular: mycorrhiza), which is Greek for "fungus roots."

The word mycorrhizae is also used to refer to the helpful fungi in fungus-root associations. The hyphae of these helpful fungi spread out into the soil in all directions, increasing their host's ability to gather nutrients by ten times or more. The fungus-root associations protect the plant against drought, cold, acid rain, and root diseases caused by harmful fungi. Currently, Australian researchers are trying to alter the hereditary material of one kind of mycorrhiza so that it produces natural insecticides. This helpful fungus would then be able to help guard its host against harmful insects as well as provide all of its normal benefits.

Lichens

Suppose someone asked you what kind of organism can live in the hot, dry desert as well as the frozen Arctic. What if the person added that this organism can also survive on bare rocks, wooden poles, the sides of trees, and even the tops of mountains? You might reply that no one organism can survive in so many different environments. In a way, your response would be right. For although **lichens** (LIGH-kuhnz) can actually live in all of these

ACTIVITY
WRITING

Taking a "Lichen" to It

Lichens are used in a variety of ways. Some are used to make dyes. Others are a source of food for people and livestock. Go to the library and find out the details about some of the ways lichens are used around the world. Prepare a report on your findings.

Figure 4–13 *Lichens show three basic patterns of growth: flat and crusty, bushy, and leafy. The British soldier lichen is a bushy lichen (right). What growth pattern do these yellow lichens show?*

environments, they are not one organism but two. **A lichen is made up of a fungus and an alga that live together.** An alga (AL-gah; plural: algae, AL-jee) is a simple plantlike autotroph that uses sunlight to produce its food. An alga lacks true roots, stems, and leaves. Blue-green bacteria, certain protists, and a number of simple plants are considered to be algae. Combined, the two organisms (alga and fungus) in a lichen can live in many places that neither could survive in alone.

The fungus part of the lichen provides the alga with water and minerals that the fungus absorbs from whatever the lichen is growing on. The alga part of the lichen uses the minerals and water to make food for the fungus and itself. Why is the relationship between the fungus and alga considered to be one of the best examples of symbiosis?

Lichens are sometimes known as pioneers because they are often one of the first living things to appear in rocky, barren areas. Lichens release acids that break down rock and cause it to crack. Dust and dead lichens fill the cracks, which eventually become fertile places for other organisms to grow. In time, the rocky area may become a lush, green forest.

4–3 Section Review

1. Discuss three ways in which fungi harm humans.
2. How are fungal diseases useful to people?
3. Explain why fungus spores are sometimes mixed into the soil in which young trees are to be grown.
4. What is a lichen?
5. Why do agricultural inspectors check samples of grain with a microscope before permitting the grain to be ground up for food?

Connection—*Ecology*

6. Many lichens are extremely sensitive to pollution. Even when there is very little pollution, these lichens will grow poorly or not at all. How might environmental scientists use this information in their work?

Murderous Mushrooms

To avoid becoming one of the several hundred cases of mushroom poisoning in the United States each year (or worse yet, a "dear departed"), a mushroom hunter must be able to accurately identify mushrooms. The difficulty of this task is particularly well-illustrated by the mushrooms in the genus *Amanita*. *A. caesarae* was a favorite food of the Caesars, the rulers of ancient Rome. *A. rubescens* is also edible and delicious—but it is poisonous when it is not sufficiently cooked. *A. muscaria* is a poisonous mushroom. In fact, people in northern Europe once soaked *A. muscaria* caps in milk and set them out to attract and kill flies. If accidentally eaten by humans, *A. muscaria* causes hallucinations, sweating, wildly crazy behavior, and deep sleep. *A. verna,* which is known by the common names fool's mushroom and destroying angel, causes cramps, severe abdominal pain, vomiting, diarrhea, liver and kidney failure, and death in over half its victims.

Interestingly, the deadly poisonous *Amanita* mushrooms taste good. (Most poisonous substances are bitter or otherwise unpleasant.) This characteristic has caused these mushrooms to be used by murderers on more than one occasion in *history.* In 54 AD, the emperor Claudius I of Rome was fed poisonous mushrooms by his wife, who wanted her son Nero to become emperor and did not want to wait for Claudius to die of natural causes. Pope Clement VII, who died in 1534, was also a victim of poisonous mushrooms.

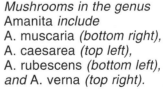

Mushrooms in the genus Amanita *include* A. muscaria *(bottom right),* A. caesarea *(top left),* A. rubescens *(bottom left), and* A. verna *(top right).*

Laboratory Investigation

Growing Mold

Problem

How does mold grow?

Materials (per group)

> large covered container
> paper towel
> piece of bread
> piece of cheese
> piece of apple
> magnifying glass

Procedure 🔺

1. Line the container with moist paper towels.

2. Place the pieces of food in the container.

3. Allow the container to remain uncovered overnight. Then cover the container.

4. Store the container in a dark place. Examine its contents daily for mold.

5. After the container's contents become moldy, examine the mold with a magnifying glass. **CAUTION:** *Be careful not to accidentally inhale mold spores. Some people are allergic to mold.*

6. Make a drawing of what you observe. Record your observations next to your drawing.

7. Observe and draw the mold every two to three days for a week.

Observations

1. When did mold start to grow?

2. How do the molds appear to the unaided eye? Under a magnifying glass?

3. How do the molds change over time?

4. Did you observe fruiting bodies and spores? What did these look like?

Analysis and Conclusions

1. Do different kinds of mold grow on different kinds of substances? Why do you think this is so?

2. On which substance did the molds grow best? Develop a hypothesis to explain why. How might you test your hypothesis?

3. Why did you leave the container uncovered? What would have happened if you had covered the container immediately?

4. **On Your Own** Design an experiment to find out how light, temperature, moisture, dust, or household disinfectants affect the growth of molds. If you receive the proper permission, you may perform the experiment you have designed.

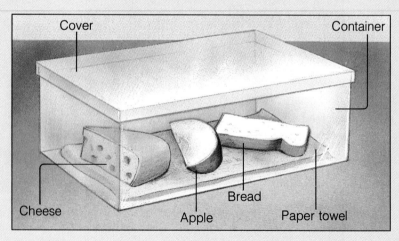

Summarizing Key Concepts

4–1 Characteristics of Fungi

▲ Fungi come in a variety of sizes, shapes, and colors.

▲ Fungi are heterotrophs. They obtain food by releasing chemicals that digest the substance on which they are growing and then by absorbing the digested food.

▲ Some fungi catch and eat tiny animals. Others are decomposers. Still others obtain food through symbiotic relationships. Some symbiotic fungi are parasites. Others help their host.

▲ Some fungi are unicellular. Most fungi are multicellular.

▲ Multicellular fungi are made up of threadlike tubes called hyphae.

▲ Fungi typically reproduce by means of spores, which are produced in structures known as fruiting bodies.

4–2 Forms of Fungi

▲ There are three basic forms of fungi: mushrooms, yeasts, and molds.

▲ Mushrooms are shaped like umbrellas. Some are good to eat. Others are poisonous.

▲ Yeasts are unicellular. They are used by bakers, brewers, industrial chemists, and medical researchers, to name a few.

▲ Molds are fuzzy, shapeless, fairly flat fungi that grow on the surface of an object. Some are used to make foods. A few are the source of important medicines.

4–3 How Fungi Affect Other Organisms

▲ Many diseases of crop and garden plants are caused by fungi.

▲ Some fungi that infect plants produce toxins that are harmful to humans and animals.

▲ Fungi cause a number of diseases in animals and humans. Some of these are simply annoying. Others are serious and even deadly.

▲ Certain fungi form fungus-root associations with most of the Earth's plants. These symbiotic relationships are extremely helpful to the plants.

▲ Lichens are produced by a symbiosis between a fungus and an alga.

Reviewing Key Terms

Define each term in a complete sentence.

4–1 Characteristics of Fungi
fungus
hypha
spore

4–2 Forms of Fungi
mushroom
yeast
mold

4–3 How Fungi Affect Other Organisms
lichen

Chapter Review

Content Review

Multiple Choice

Choose the letter of the answer that best completes each statement.

1. Most fungi reproduce by means of
   a. budding.
   b. spores.
   c. binary fission.
   d. conjugation.
2. Yeasts consist of
   a. many long hyphae that are tightly wound together.
   b. a stemlike stalk that is topped by a cap.
   c. single cells.
   d. cup-shaped fruiting bodies.
3. Fungi that get their food from the remains of dead organisms are known as
   a. decomposers.
   b. producers.
   c. parasites.
   d. symbionts.
4. Fungi produce spores in structures known as
   a. buds.
   b. stalks.
   c. puffballs.
   d. fruiting bodies.
5. Which of the following is not necessary for fungi to grow and survive?
   a. sunlight
   b. proper temperature
   c. moisture
   d. food

6. A lichen is a symbiotic partnership between a
   a. yeast and a mold.
   b. fungus and a plant's roots.
   c. mold and an ant.
   d. fungus and an alga.
7. Which of the following is not a disease caused by fungi?
   a. ringworm
   b. mycorrhiza
   c. chestnut blight
   d. ergot poisoning
8. Any close relationship between two organisms in which at least one of the organisms benefits is known as
   a. heterotrophy.
   b. symbiosis.
   c. autotrophy.
   d. parasitism.
9. The month-old leftovers that you discover at the back of the refrigerator are spotted with a velvety blue-green fungus. This fungus is probably a
   a. yeast.
   b. mushroom.
   c. lichen.
   d. mold.

True or False

If the statement is true, write "true." If it is false, change the underlined word or words to make the statement true.

1. Most fungi are made up of threadlike structures known as <u>fruiting bodies</u>.
2. The scientific name for the organisms called <u>lichens</u> is mycorrhizae.
3. Fleming discovered that a certain mold produced an antibiotic that he called <u>cyclosporine</u>.
4. The three basic forms of fungi are <u>mushrooms, molds, and mildews</u>.
5. Fungi are classified according to <u>characteristics that show their evolutionary relationships</u>.

Concept Mapping

Complete the following concept map for Section 4–1. Refer to pages B6–B7 to construct a concept map for the entire chapter.

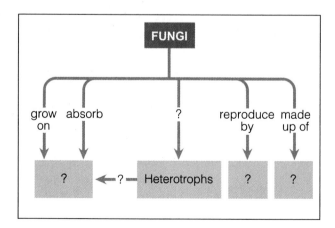

Concept Mastery

Discuss each of the following in a brief paragraph.

1. Explain why fungi are placed in their own kingdom.
2. How do fungi obtain energy and nutrients?
3. How do the alga and the fungus in a lichen help each other?
4. Describe three ways in which molds help humans.
5. Why are yeast cells used in baking bread?
6. Describe the structure of a typical mushroom and explain the function of each structure.
7. Explain how fungi that infect plants can harm humans.

Critical Thinking and Problem Solving

Use the skills you have developed in this chapter to answer each of the following.

1. **Relating concepts** What is the connection between reproduction by spores and the rapid spread of crop diseases caused by fungi?
2. **Designing an experiment** Design a set of experiments to show how light, temperature, and dryness affect the growth of bread mold. Be sure to include a control in each of your experiments—and make sure that each experiment tests only one variable.
3. **Making inferences** As you can see in the accompanying photograph, a sharpei dog looks like its skin is many sizes too large. People who own a sharpei dog must rearrange its loose folds of skin every now and then. This is especially true in areas where there is a lot of moisture in the air. What might happen to a sharpei dog if its owner neglected to rearrange its folds of skin?
4. **Assessing concepts** Fungi are sometimes divided into two groups: yeasts (unicellular fungi) and molds (multicellular fungi). Is this informal classification system useful? Why or why not?
5. **Using the writing process** While looking something up in the local library's new set of encyclopedias, you notice that fungi are defined as "plants that lack chlorophyll (the green substance involved in food-making)." Write a letter to the editors of the encyclopedia explaining why this is not a good description of fungi.
6. **Using the writing process** Imagine that you are the alga in a lichen. Write a letter to an old friend describing your fungal partner.

Plants Without Seeds

The waves lapping around your face are so cold that your skin goes numb. You make a few last-minute adjustments to your wet suit, then raise your gloved hand, forming a circle with your thumb and forefinger—O.K. One of your friends on the boat gives you a thumbs-up. With a practiced movement of your arms, you descend feet first into an alien world.

Once underwater, you find yourself in a strange, dreamlike forest. Thin, vinelike "trees" stretch up toward the silver sky far above your head. Below you, the trees extend so far down into murky darkness that they seem to go on forever. Fishes dart like birds among the swaying ribbonlike leaves of the trees. A curious young harbor seal suddenly appears and stares at you with large dark eyes. Then it does a flip with a half-twist and vanishes like a ghost among the long, thin trees of seaweed.

The trees of this underwater forest are a type of seaweed known as kelp. Kelp forests—found off the coasts of Washington, Oregon, and northern California—are home to slugs, snails, fishes, seals, sea otters, and many other living things.

Kelps are just one kind of plant without seeds. In this chapter, you will read about the strange and ancient world of plants without seeds.

Journal *Activity*

You and Your World In your journal, record your thoughts about studying plants. Do you think this chapter and the one that follows will be easy? Difficult? Interesting? When you have completed Chapters 5 and 6, look back at this entry and see if the chapters were as you expected.

◀ *Long, ropy strands of kelp sway gracefully with the movement of the ocean's waves and currents as a diver explores the mysterious, dimly lit world of the seaweed "forest."*

Guide for Reading

Focus on this question as you read.

▶ *How are red, brown, and green algae similar? How are they different?*

5–1 Plants Appear: Multicellular Algae

Have you ever walked along a beach? If so, you may have found flat greenish-brown ribbons, brown ropes, or delicate reddish tangles of seaweed washed up on the sand. You may have discovered bits of yellowish seaweed that contain bubbles which squish with a satisfying pop. Looking into tide pools, you might have seen transparent green veils of sea lettuce, dainty mermaid's cups, or pink coralline algae.

Even if you have never been to a beach, you have probably encountered seaweed in other forms. For example, the edible blackish wrapper on certain kinds of sushi (Japanese rice rolls) is made of dried seaweed. Ground-up seaweed is mixed with water and sprayed on gardens to make plants grow better. And chemicals from seaweed are added to ice cream, jellies, candy, and many other foods to give them a smooth texture.

Seaweeds are some of the most familiar types of multicellular **algae** (AL-jee; singular: alga, AL-gah). As you learned in Chapter 4, the term alga refers to any simple plantlike autotroph that uses sunlight to produce its food. Thus the term algae is used to refer to everything from microscopic unicellular blue-green bacteria to enormous multicellular strands of kelp as long as a football field.

Scientists do not agree about the formal classification of algae. Some scientists think that all algae (except blue-green bacteria) should be classified as protists. Others think that all algae should be classified

Figure 5–1 *Algae come in many forms. Mermaid's cups resemble shallow drinking glasses (top). The bubblelike structures on giant kelp help to keep it afloat in its watery home (bottom right). The delicate branches of coralline algae contain so much limestone that they look and feel as if they were carved out of reddish stone (bottom left).*

as plants. Both of these views have evidence to support them. In this textbook, we have decided to use the following classification system: Blue-green bacteria belong to the kingdom Monera. Most of the other unicellular algae are assigned to the kingdom Protista. Multicellular algae and closely related species of unicellular algae are placed in the kingdom Plantae.

In Chapter 2, you learned about blue-green bacteria. In Chapter 3, you learned about plantlike protists. In this chapter, you will learn about multicellular algae and the unicellular algae that are closely related to them. For convenience and simplicity, from this point on we will refer to multicellular algae and their close relatives as algae.

Algae were the first kinds of plants to appear on Earth. The oldest fossils (preserved remains of ancient organisms) of algae are about 900 million years old. Fossil evidence indicates that land plants evolved from certain types of these ancient algae.

Although algae resemble the land plants that are familiar to you, there are some important differences. **Algae lack the special tubes that transport water and other materials through the bodies of land plants.** This means that algae do not have true roots, stems, or leaves. By definition, roots, stems, and leaves contain these special tubes. **Algae do not have seeds.** Because they lack transporting tubes and seeds, algae must live in or near a source of water in order to survive and reproduce.

Algae are divided into three phyla: **brown algae, red algae,** and **green algae.** The phyla get their names from the **pigments,** or colored chemicals, that are found within the cells of the algae.

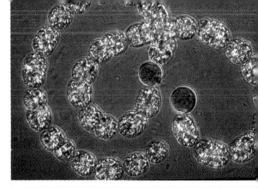

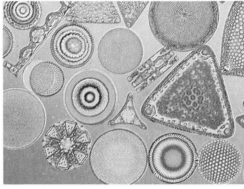

Figure 5–2 *Blue-green bacteria (top) and diatoms (bottom) are both types of algae that do not belong to the plant kingdom.*

Figure 5–3 *Multicellular algae are divided into three phyla according to the pigments they contain. Which of the algae shown here is a red alga? Which is a green alga? Which is a brown alga?*

In green algae, the most noticeable pigment is the green chemical chlorophyll. Chlorophyll captures light energy so that it can be used in the food-making process. Red algae and brown algae also contain chlorophyll. However, the green color of the chlorophyll is masked by other kinds of pigments. These pigments, which are known as accessory pigments, absorb light energy and transfer it to chlorophyll for use. The main accessory pigments in red algae are pink, red, reddish purple, and reddish black. What color would you expect the main accessory pigment in brown algae to be?

Brown Algae

For centuries it was the subject of sailors' nightmares—a haunted sea located somewhere in the Atlantic Ocean between the African coast and the islands of the West Indies. The sailors whispered the name of this sea when they dared to say it at all—the Sargasso Sea. According to legend, if a ship was foolish or unlucky enough to sail into this sea, it would encounter seaweed that could cling to the sides of the ship like monster hands. As the ship sailed on, the seaweed would become thicker and thicker, slowing the ship to a crawl, and then to a halt. A ship trapped in the horrible seaweed of the Sargasso Sea would remain there forever, joining a fleet of dead ships guarded by skeletons.

The legends about the ghost ships of the Sargasso Sea are purely imaginary. Although there really is a Sargasso Sea, it is not an obstacle to ships. In the Atlantic Ocean between Africa and Bermuda lies the real Sargasso Sea. It is an area of calm winds and gentle waves that is a perfect home for the brown algae *Sargassum* (sahr-GAS-uhm). *Sargassum* floats on or near the ocean's surface, held up by tiny round air-filled structures that act like inflatable life preservers. These air-filled structures, or air bladders, help to ensure that the *Sargassum* stays close to the surface of the ocean, where it can get plenty of sunlight. Why is sunlight important for *Sargassum*?

In most parts of the ocean, a clump of seaweed would quickly be torn apart by the action of the wind and waves. But there is little wind and wave action in the Sargasso Sea. As a result, *Sargassum* is

ACTIVITY
DOING

That's About the Size of It

1. Using a meterstick, determine the height in meters of each person in your science class (including your teacher). Record your measurements.

2. Using a calculator, add up the heights of all the people in your class (make sure you put the decimal point in the right place!). You have now defined the length of an unofficial unit of measurement known as a bakalian. How many meters long is a bakalian?

3. Giant kelps can be as long as 100 meters. How does the length of a bakalian compare to that of such a kelp plant? How long is a 100-meter kelp plant in bakalians?

able to form enormous floating mats many kilometers long. These mats are home to fishes, crabs, jellyfish, and many other ocean animals. Can you find the crab and the fish hiding among the strands of *Sargassum* in Figure 5–4?

While the Sargasso Sea did not provide an answer to the mysterious disappearance of sailing ships, it did explain a few biological mysteries. One of these mysteries involved the snakelike fishes known as eels. For more than two thousand years, people wondered where eels in the lakes and rivers of Europe came from. No one had ever seen a baby eel. No one had ever seen eels mate or lay eggs. Aristotle, a philosopher and scientist of ancient Greece, thought that eels simply sprang out of the mud. During the eighteenth century, some people thought that eels were formed from the hairs on horses' tails.

It was not until the end of the nineteenth century that the mystery was solved: Adult eels leave the rivers and lakes and travel thousands of kilometers to the Sargasso Sea to mate and lay their eggs. The baby eels, which are leaf-shaped fish that look nothing like adult eels, live for a while among the mats of *Sargassum*. Then, over a period of several years, the baby eels gradually travel back across the ocean to Europe. As they grow, the baby eels slowly change their shape. By the time they reach the mouths of the European rivers and begin swimming upstream, the baby eels have grown up into small adult eels.

Sargassum is not the only kind of brown algae. The kelp that you read about earlier is another type of brown algae. So is the rockweed that grows on rocky coasts. As you can see in Figure 5–6 on page 110, brown algae come in a wide variety of shapes

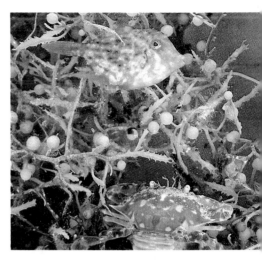

Figure 5–4 *The enormous floating mats of* Sargassum *are home to many animals. The spherical structures on the* Sargassum *are air bladders. What is the function of the air bladders?*

Figure 5–5 *Baby eels spend the first few years of their lives in the Sargasso Sea. Then they travel to the rivers of Europe. On their journey, they take on their adult shape (left) and develop pigments so they are no longer transparent. Silvery-gray adult European eels may wiggle across patches of dry land as they begin their journey back to the Sargasso Sea to breed (right).*

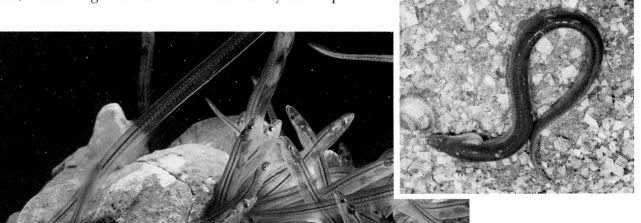

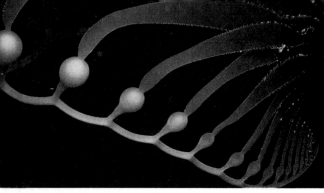

Figure 5–6 *Brown algae come in all shapes and sizes. Can you locate the air bladders on the kelp (top right)?*

and sizes. Most brown algae live in the ocean. Some, like *Sargassum*, float freely. Others, such as kelps and rockweed, are attached to the sea floor.

Brown algae have long been used as food for humans in China, Japan, Canada, Ireland, New Zealand, and many other parts of the world. They can also be used as food for livestock. For example, in some places in Scotland and Ireland, cattle and sheep graze on brown algae at low tide. Chemicals extracted from brown algae (known as algins or alginates) are added to salad dressings and other foods to make them smooth and prevent their ingredients from separating.

Red Algae

Like brown algae, most red algae are multicellular and live in the ocean. Red algae can grow to be several meters long, but they never reach the size of brown algae such as *Sargassum* or giant kelps. The shapes of red algae are just as varied as those of brown algae. Some red algae form clumps of delicate, branching red threads. Others grow in large, rounded, flat sheets. Still others produce hard, stiff branches rich in calcium carbonate (the substance that makes up the shells of foraminifers, corals, snails, and crabs).

Red algae usually grow attached to rocks on the ocean floor. Red algae can be found at depths up to 170 meters—far deeper than other kinds of algae. Very little sunlight penetrates to these extreme depths. Chlorophyll alone cannot absorb enough light energy to allow the food-making process to continue at a life-sustaining level. How, then, do deep-water red algae get the light they need to survive?

The answer involves the accessory pigments that give red algae their characteristic color and their

name. The accessory pigments are able to absorb the small amount of light energy that penetrates to deep waters. The absorbed energy is then transferred to chlorophyll. Thus the accessory pigments enable red algae in deep water to make the most of the light that reaches them. Now can you explain why red algae are able to live in deeper waters than green algae?

Red algae are used by humans in a number of ways. Some species are eaten as food. Other species are harvested by the ton and added to soil to make it better for crops. Chemicals extracted from red algae are used to manufacture certain foods. The next time you eat ice cream, use a creamy salad dressing, drink chocolate milk, or frost a cake with ready-made frosting, read the list of ingredients on the package. You may find carrageenan (KAIR-uh-geen-uhn), a substance that comes from red algae, on the list. Another substance derived from red algae is agar (AH-gahr). Agar plays an important role in medical research. It is used to make the jellylike nutrient mixtures on which bacteria and other microorganisms are grown.

Figure 5–7 *Red algae may resemble fingers of transparent red cellophane, dark pink rock formations, or clumps of purplish-black hair.*

Activity Bank

Seaweed Sweets, p. 180

Green Algae

Deep in space, a silvery ship is in its second year of a four-year journey to study the planets. On board, the astronauts are about to finish dinner. Although they have brought along enough food to last the entire journey, they could not carry enough oxygen to last several years. Are the astronauts doomed to suffocate in space? Of course not.

In a tiny room near the back of the ship sits a tank of water filled with green algae. The green algae use the carbon dioxide exhaled by the astronauts

Figure 5–8 *The white streaks on the agar in this petri dish consist of millions of bacteria. Because most microorganisms cannot live on agar alone, nutrients and other substances are added to the agar. In this case, blood has changed the color of the agar from pale yellow to red.*

Figure 5–9 *Some green algae look like exotic vines with cone-shaped leaves. Others form enormous green bubbles.*

Figure 5–10 *Hydras are freshwater relatives of jellyfishes and sea anemones. The greenish color of* Chlorohydra *is caused by the hundreds of green algal cells that live within its body. How do the algal cells help their partner?*

to make food. During the food-making process, the algae release oxygen, which the astronauts breathe. If you could look closely at the tank, you would see bubbles of oxygen floating toward the surface. The relationship between the breathing of the astronauts and the food-making process in the algae means that the astronauts and the green algae support each other's lives.

This scene is purely imaginary. However, science fiction may soon become science fact. Today, scientists are hard at work developing methods of growing green algae in closed environments. And someday, future space explorers may rely on algae to maintain the air supply aboard spacecraft.

For now, the only place you can find green algae is here on planet Earth. Most green algae live in fresh water or in moist areas on land. However, some green algae do live in rather unusual places. Several may live in a close partnership with fungi, forming lichens. A few can live in bodies of water many times saltier than the ocean, such as the Great Salt Lake in Utah. And some can live within the bodies of worms, sponges, and protists such as *Paramecium.*

Green algae have life cycles, pigments, and stored food supplies similar to those of complex land plants. These similarities, along with a few others, suggest that land plants evolved from green algae. Of course, modern species of green algae are not the ancestors of land plants. But in the distant past, modern species of land plants and modern species of green algae shared a common ancestor. Because the remains of algae rarely become fossils, it is unlikely that we will ever know exactly what this common ancestor was. But by examining modern species of green algae and using imagination, scientists have developed a pretty good idea of how land plants may have evolved from green algae.

Modern species of green algae that represent important steps in the development of land plants are shown in Figure 5–11. As you can see, the earliest forms of green algae were probably unicellular organisms. Later, colonies consisting of many relatively independent cells developed. The next step on the way to land plants were multicellular green algae that lived in water. After that came multicellular green algae that lived on land. Finally, about 450 to 500 million years ago, algaelike land plants evolved from green algae ancestors. One group of early land plants evolved into mosses and their relatives. Another group evolved into ferns and other more complex land plants.

Figure 5–11 *By studying modern forms of green algae, scientists have been able to reconstruct the basic stages of the evolution of plants. Unicellular algae, such as* Chlamydomonas *(top left), gave rise to colonial algae such as* Volvox *(bottom left). Colonial algae gave rise to simple multicellular algae such as* Ulva *(top right), which in turn gave rise to more complex multicellular forms (bottom right). What was the next big step in the evolution of land plants from algae?*

5–1 Section Review

1. Compare the three phyla of algae.
2. Why can red algae survive in deeper water than other kinds of algae can?
3. Why are brown algae important to humans?
4. How are green algae related to land plants?

Connection—*Social Studies*

5. On his first voyage to the New World, Christopher Columbus had a lot of trouble with sailors who wanted to turn back. Some historians think that the sailors were afraid of falling off the edge of the world. Other historians disagree. Explain how brown algae may have played a part in the near-mutiny of Columbus's sailors.

Guide for Reading

Focus on these questions as you read.

▶ What are some adaptations that plants need to live on land?

▶ What are the major characteristics of mosses and their relatives?

5–2 Plants Move Onto Land: Mosses, Liverworts, and Hornworts

In the ocean, a kelp plant may stand as tall as a very large tree. Its leaves are held up by the water so that they are exposed to enough light. Kelp absorbs water, carbon dioxide, minerals, and all the other substances it needs from ocean water. It reproduces by releasing egg cells and sperm cells into the water, where the sperm cells can swim to the eggs and fertilize them.

Now imagine that the same kelp plant is taken from the water and planted on land. Can the plant stand upright and hold up its leaves? Can it absorb minerals and water from the new substance (air) that surrounds it? Can its sperm cells swim on dry land? Can the kelp plant survive on land? The answer to all of these questions is no.

As the example of the kelp plant indicates, it is not easy for aquatic organisms to live on dry land. The plants that invaded the land millions of years

Figure 5–12 *The first forests appeared on Earth more than 380 million years ago. Unlike the forests of today, these ancient forests were dominated by plants without seeds. What adaptations did plants require in order to successfully invade the land?*

ago had to evolve in ways that made them better suited to meet the challenges of their new environment. Let's look at some of the tasks that plants need to perform in order to survive on land.

Land plants need to support the leaves and other parts of the body so that they do not collapse. Supporting structures enable land plants to position the parts of their body that make food so that they are exposed to as much sunlight as possible.

Land plants need to obtain water and minerals. In an aquatic environment, plants are surrounded by water and dissolved minerals. On land, however, water and minerals are located in the soil.

Land plants need to transport food, water, minerals, and other materials from one part of the body to another. In general, water and minerals are taken up by the bottom part of a land plant and food is made in the top part. To supply all the cells of the body with the substances they need, water and minerals must be transported to the top part of the plant and food must be transported to the bottom part.

Land plants need to prevent excess water loss to the environment. Because air contains less water

than body cells, the body cells of land organisms tend to lose water. (Have you ever noticed how a puddle of rainwater on a sidewalk gradually shrinks and then disappears? This happens because there is a greater concentration of water in the puddle than in the air, and the water evaporates. The same basic principle applies to living cells.) To avoid drying up, land organisms need ways of minimizing water loss.

Land plants need to get sperm cells and egg cells together so that reproduction can occur. Sperm cells need water in order to swim and to fertilize egg cells.

Now that you know what the necessary tasks are, let's look at the ways in which mosses, liverworts, and hornworts accomplish these tasks. As you read about mosses and their relatives, you may want to refer to Figure 5–13 to see what these plants look like.

Mosses, liverworts, and hornworts are tiny plants that live in moist places. They can be found on wet rocks, damp tree bark, and the muddy banks of ponds and streams. In some places, such as bogs and forests, they may cover the ground like a fuzzy green carpet. Because they are small and live in places where water is plentiful, they have few special adaptations for dealing with the challenges of life on

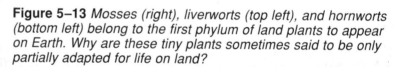

Figure 5–13 *Mosses (right), liverworts (top left), and hornworts (bottom left) belong to the first phylum of land plants to appear on Earth. Why are these tiny plants sometimes said to be only partially adapted for life on land?*

Figure 5–14 *At one stage in their life cycles, mosses produce many dustlike spores. If a spore lands in a place with favorable growing conditions, it will grow into a new leafy green moss plant.*

land. (This is quite different from the more complex plants you will read about later, which have many adaptations for survival on land.) You can think of mosses and their relatives as representing one solution to the problems of living on land: avoiding the most difficult challenges.

Because they are small and low to the ground, these plants do not need sturdy stems or other special supporting structures. The stiff, rigid cell walls that surround their cells provide all the support they need. Their small size also means that they do not need a special system to transport materials throughout their body. They can simply transfer materials from one cell to the next. This method of transport does not allow materials to be carried very far or very efficiently—but it is good enough for a very small plant.

Because the places in which they live are quite moist, mosses, liverworts, and hornworts do not need an adaptation to prevent water loss, such as a waxy covering. However, the reproductive cells that develop into new plants often have a thick, watertight coat. This enables the reproductive cells to survive dry periods. The moist conditions in their environment make it possible for mosses and their relatives to carry out certain life functions as if they lived in water. For example, these plants absorb water and minerals through their entire body surface. And their sperm cells swim to the egg cells when the body of the plant is covered with rainwater or dew.

ACTIVITY

THINKING

Moss-Grown Expressions

Consider these two old sayings:

1. A rolling stone gathers no moss.

2. Moss grows on the north side of a tree.

Do you think there is any truth to either of these sayings? Why or why not?

Figure 5–15 *In Washington State's Olympic National Park, mosses thrive in the cool, extremely damp Hoh Rain Forest. There, the mosses cover the trees like shaggy green coats (top). In most other forests, mosses do not cover the trees but may form a soft carpet on the forest floor (bottom).*

Although mosses are not particularly impressive plants, they are quite useful to humans. Certain Japanese-style gardens feature gently curving mounds of soil covered with a plush layer of emerald green moss. At least one modern artist has created living patchwork quilts of mosses. Dried sphagnum (SFAG-nuhm) moss is added to garden soil to enrich the soil and improve its ability to retain water. Sphagnum moss also changes the chemical balance in the soil so that the soil is better for growing certain plants, such as azaleas and rhododendrons. Ground-up sphagnum moss is added to soil and sprinkled around seedlings to help prevent the growth of certain disease-causing bacteria and fungi. In the past, people have used sphagnum moss to treat burns and bruises and to bandage wounds. Can you explain why sphagnum moss may have been a good substance for covering a wound?

Over the course of hundreds of years, layer after layer of sphagnum moss and other plants may accumulate at the bottom of a bog or swamp. Under the right conditions, the deposits of dead plants form a substance called peat. Peat is used as a soil conditioner. It is also dried and used as fuel. Because the conditions in peat bogs greatly slow the process of decay, scientists who study ancient civilizations have found many interesting things within the deposits of peat—including the lifelike body of a man who was buried about 2000 years ago.

5–2 Section Review

1. List five tasks that plants need to perform in order to live on land.
2. How are mosses and their relatives adapted for life on land?
3. What are three ways in which humans use mosses?

Critical Thinking—*Relating Cause and Effect*
4. Explain why mosses and their relatives never grow very large.

5–3 Vascular Plants Develop: Ferns

Guide for Reading

Focus on this question as you read.

▶ How are ferns different from other kinds of plants without seeds?

How would you like to be able to make yourself invisible? Sound like fun? Well, folklore has it that all you have to do is gather fern seed by the light of the moon at midnight on Saint John's Eve. Carrying some fern seed so gathered is guaranteed to make you invisible to human eyes.

But before you go off into the woods on a mid-summer night to look for ferns, you should know there is a catch. Like the algae, mosses, liverworts, and hornworts you read about earlier, ferns do not have seeds.

Unlike the other seedless plants you read about in this chapter, however, ferns do have a number of special adaptations to life on land. For example, they have a waxy covering on their leaves that helps to prevent water loss and roots that enable them to gather water and minerals from the soil. The most important adaptations, however, involve a system of tiny tubes that transport food, water, and other materials throughout the body of the fern. These tiny tubes are known as vascular tissue. **Because they have vascular tissue, ferns are said to be vascular plants.** Plants that lack vascular tissue—such as algae and mosses—are known as **nonvascular plants.**

The first **vascular plants,** which appeared about 400 million years ago, represent a major step in plant evolution. Vascular plants, such as ferns, are much better adapted to life on land than nonvascular plants. Vascular tissue allows materials to be

Figure 5–16 Ferns do not have seeds. The dots or lines visible on the underside of fern leaves are formed by clusters of spore cases (left). As this 200-million-year-old fossil indicates, ferns are among the most ancient types of vascular plants (right).

Figure 5–17 *Vascular plants without seeds include horsetails (left) and club mosses such as the peacock "fern" (right). They also include ferns.*

Figure 5–18 *A close examination of this tree fern leaf reveals that it is divided into smaller leaflike parts, which in turn are divided into still smaller leaflike parts. A few ferns have leaves that are not at all fernlike. For example, the water fern* Marsilea *has rounded leaves that make it look like a four-leaf clover.*

transported quickly and effectively throughout the body of a plant. In addition, the type of vascular tissue that carries water from the roots in the soil to the leaves in the air is made of cells that have extremely thick, strong cell walls. These sturdy cell walls greatly strengthen the stems. Now can you explain why vascular plants can grow much larger than nonvascular plants?

Thanks to their vascular tissue, ferns can grow much taller than mosses and their relatives. In the mainland United States, most ferns range in height from a few centimeters to about a meter. Tree ferns, which grow in rain forests in Hawaii, the Caribbean, and other tropical areas, can be enormous. If laid along the ground, the tallest tree fern would extend from the pitcher to the batter on a baseball diamond, or from the foul line to the pins in a bowling lane.

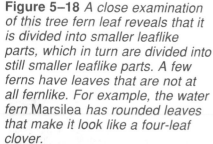

Like other vascular plants, ferns have true leaves, stems, and roots. As you can see in Figure 5–18, the leaf of a fern is often divided into many smaller parts that look like miniature leaves. In many types of ferns, the developing leaves are curled at the top and look much like the top of a violin. Because of their appearance, these developing leaves are called fiddleheads. As they mature, the fiddleheads uncurl until they reach their full size.

People often mistake the sticklike central portion of a fern leaf for a stem. This is quite understandable when you realize that the stems of most ferns are hidden from view. In some ferns, the stems look like fuzzy brown strands of yarn lying on the ground. The stems are hidden by the feathery fern leaves that emerge from them. In other ferns, the stems run below the surface of the soil.

If you were to pull part of a fern's stem from the ground, you would discover clumps of wiry structures growing down from the stem here and there. These structures are the fern's roots. The roots anchor the fern to the ground. They also absorb water and minerals for the plant.

Although ferns are much better adapted to life on land than other nonvascular plants, they are not fully adapted. Take a moment now to look back at the list of tasks that plants need to perform in order to survive on land (pages 115–116). How do ferns accomplish each task? Can you guess for which task ferns require a watery environment? That's right: Ferns need standing water in order for their sperm cells to swim to their egg cells. Like other plants without seeds, they need an abundant supply of water in order to reproduce.

Ferns are useful to humans in a number of ways. Because ferns have lovely, interestingly shaped leaves and thrive in places that have little sunlight, they are popular houseplants. Products from ferns are often used to grow other kinds of houseplants. For example, orchids may be grown on the tangled masses of fern roots or on fibrous chunks of tree-fern stems. Ferns may also assist in the growth of crops. In southeast Asia, farmers grow a small aquatic fern in rice fields. Tiny pockets in the fern's leaves provide a home for special blue-green bacteria that produce a natural fertilizer. Thus the fern and its microscopic

Figure 5–19 *The developing leaves of most ferns are tightly coiled. This makes them look like the scroll at the top of a violin. What are these developing fern leaves called?*

ACTIVITY
READING

The Secret of the Red Fern

What would you expect a book called *Where the Red Fern Grows* to be about? No, this wonderful book by Wilson Rawls is not about ferns and where they can be found. However, the red fern plays a small but important part in this story. What is the meaning of the red fern? Why does the author include it in this story? Read, and find out.

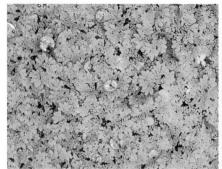

Figure 5–20 *Ferns are valued ornamental plants. The* Dipteris *fern has paired leaves that look like lacy wings (top right). The leaves of maidenhair ferns are divided into many small, dainty leaflike parts (bottom right). A young leaf from a Sri Lankan fern is a beautiful red color when it first uncoils (bottom left). Ferns may also be valued for reasons other than their beauty. For example, the tiny water fern* Azolla *(top left) helps to fertilize rice fields.*

partner help the farmers to grow food. Some ferns are eaten directly as food. During the spring, edible (fit to be eaten) fern fiddleheads may be sold in specialty food shops, supermarkets, and roadside vegetable stands. When properly cooked, fiddleheads make a delicious vegetable dish. However, unless you are absolutely certain which ferns are edible, you should not gather fiddleheads for food.

5–3 Section Review

1. What is the most important difference between ferns and the other plants without seeds that you studied in previous sections?
2. Describe the structure of a typical fern.
3. How are ferns adapted to life on land?

Critical Thinking—*Designing an Experiment*
4. Design a series of experiments to determine the best conditions for growing ferns.

I Scream, You Scream, We All Scream for Ice Cream

You know that *ice cream* is sweet, cold, soft, and creamy—a wonderful treat. But did you know that you can learn a lot about *physical science* by studying ice cream? That's right—ice cream can be the key to understanding many scientific principles.

Consider for a moment the stuff ice cream is made of: milk and cream, sugar, a dash of flavoring, and perhaps a pinch of agar or gelatin. Blended together, these ingredients form a syrupy liquid. This liquid mix is transformed into a solid dessert by removing heat energy—that is, by freezing it. And this illustrates an important science concept: The form, or phase, of matter can be changed by adding or taking away energy.

But heat energy does not simply vanish. It has to go somewhere. (Scientists know this principle as the first law of thermodynamics.) In an old-fashioned ice-cream maker, the heat energy is taken up by a mixture of ice and rock salt. Because nature tends to work to even things out (scientists call this principle the second law of thermodynamics), heat energy is transferred from the ice-cream mix, which has more heat energy

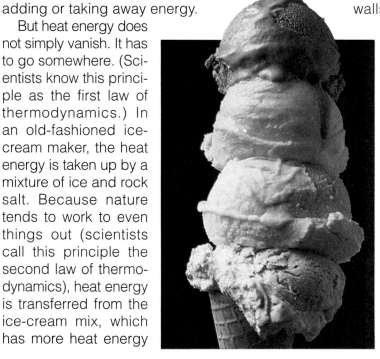

to the mixture of ice and rock salt, which has less heat energy.

Have you ever tried to restore a dish of melted ice cream by putting it into the freezer? If you have, you know that the resulting frozen substance is hard and unappetizing—more like ice than ice cream. Why? One of the most important ingredients of ice cream is air. Turning the crank on an old-fashioned ice-cream maker stirs air into the mix (as well as helps the mix cool evenly). Tiny bubbles of air about 0.1 mm across make ice cream soft. The need for bubbles is exactly the reason agar, carrageenan, or gelatin is often added to ice-cream mixes. These substances help to stabilize the walls of the air bubbles. The air bubbles are also the reason ice cream is not served on airplanes. You see, the air pressure inside an airplane flying high above the ground is much lower than the air pressure on the ground. At these lower pressures, air escapes from the ice cream. So a quart of soft, delicious ice cream can easily be turned into a pint of hard, not-so-tasty ice cream—hardly a wonderful treat!

Laboratory Investigation

Comparing Algae, Mosses, and Ferns

Problem

How are algae, mosses, and ferns similar? How are they different?

Materials (per group)

```
3 microscope slides
3 coverslips
medicine dropper
microscope
hand lens
metric ruler
scissors
brown alga plant
moss plant
fern plant
```

Procedure 🧪 ▭

1. Examine the brown alga carefully. Use the hand lens to get a closer look at the plant.

2. Draw a diagram of the alga on a sheet of unlined paper. Try to be as neat, accurate, and realistic as you can.

3. Next to your diagram, write down your observations about the color, texture, flexibility, and any other characteristics of the plant that you think interesting or important.

4. Using the metric ruler, measure the length and width of the entire plant. You should also measure any parts of the plant that you think appropriate. Write these measurements on or next to the appropriate part of your diagram.

5. Using the scissors, carefully cut a piece about 5 mm long from the tip of a "leaf" on the plant.

6. Place the plant piece in the center of a glass slide. With the medicine dropper, place a drop of water on the plant piece. Cover with a coverslip.

7. Examine the plant piece under the low- and high-powers of the microscope. Draw a diagram of what you observe.

8. Examine the moss plant using the procedure outlined in steps 1 through 7.

9. Examine the fern plant using the procedure outlined in steps 1 through 7.

Observations

1. What are the dimensions of the alga, moss, and fern?

2. Which land plant grows larger—the moss or the fern?

3. Which plant has vascular tissue? How can you tell?

4. How do the top and bottom surfaces of the fern leaf differ from each other?

Analysis and Conclusions

1. Why is the fern able to grow larger than the moss?

2. What alga structures appear to be adaptations to life in water? Explain.

3. Why is one side of the fern leaf shinier than the other? How is this an adaptation to life on land?

4. How are the plants similar?

5. How are the plants different?

6. **On Your Own** Examine a liverwort, green alga, and/or red alga. How do these plants compare to the ones you examined in this investigation? How are they adapted to the places in which they live?

Study Guide

Summarizing Key Concepts

5–1 Plants Appear: Multicellular Algae

▲ Algae are nonvascular plantlike autotrophs that use sunlight to produce food. Algae do not have true roots, stems, or leaves.

▲ The algae that are classified as plants are divided into three phyla: brown algae, red algae, and green algae.

▲ Algae were the first kinds of plants to appear on Earth.

▲ Algae must live in or near a source of water in order to survive.

▲ Chlorophyll captures light energy so that it can be used in the food-making process.

▲ Accessory pigments absorb light energy and then transfer it to chlorophyll.

▲ Algae are useful to humans in many ways.

▲ Land plants evolved from green algae.

5–2 Plants Move Onto Land: Mosses, Liverworts, and Hornworts

▲ Adaptations of plants for living on land include structures for support, roots, vascular tissue, structures for minimizing water loss, and methods of reproduction that do not require standing water.

▲ Mosses, liverworts, and hornworts are small nonvascular plants that live in moist places. They are not fully adapted for life on land.

5–3 Vascular Plants Develop: Ferns

▲ Vascular plants, such as ferns, are much better adapted to life on land than nonvascular plants.

▲ Vascular tissue, a system of tiny tubes within vascular plants, allows food, water, and other materials to be transported quickly and effectively throughout the body of a plant.

▲ Water-conducting vascular tissue helps to strengthen and support stems.

▲ Vascular plants have true roots, stems, and leaves.

▲ Ferns are one of the earliest types of vascular plants. Although ferns have many adaptations to life on land, they are still dependent on standing water for reproduction.

Reviewing Key Terms

Define each term in a complete sentence.

5–1 Plants Appear: Multicellular Algae
alga
brown alga
red alga
green alga
pigment

5–3 Vascular Plants Develop: Ferns
nonvascular plant
vascular plant

Chapter Review

Content Review

Multiple Choice

Choose the letter of the answer that best completes each statement.

1. Which of these is a vascular plant?
   a. moss   c. *Sargassum*
   b. hornwort   d. fern
2. The most important characteristic in the classification of multicellular algae and their close relatives is the
   a. structure of the leaves.
   b. type of pigments present.
   c. method of producing or obtaining food.
   d. presence or absence of a nucleus.
3. The first kinds of plants to invade the land were probably
   a. algae.   c. mosses.
   b. ferns.   d. liverworts.
4. Ferns have
   a. simple seeds.
   b. no special adaptations for life on land.
   c. true roots, stems, and leaves.
   d. all of these.

5. Most agar comes from
   a. mosses.   c. red algae.
   b. ferns.   d. green algae.
6. If they existed today, the ancestors of land plants would probably be classified as
   a. ferns.   c. brown algae.
   b. red algae.   d. green algae.
7. All ferns, green algae, and mosses
   a. live only in water.
   b. are multicellular.
   c. lack vascular tissue.
   d. contain chlorophyll.
8. Mosses, liverworts, and hornworts
   a. rely on air bladders for support.
   b. can grow as tall as 60 meters.
   c. have true roots, stems, and leaves.
   d. require abundant water for reproduction.

True or False

If the statement is true, write "true." If it is false, change the underlined word or words to make the statement true.

1. *Sargassum* is a type of <u>red algae</u>.
2. Plants without seeds have methods of reproduction that <u>require</u> standing water.
3. In your textbook, <u>all</u> algae are classified as plants.
4. Mosses and their relatives have <u>few</u> special adaptations for life on land.
5. Red and brown algae <u>do not contain</u> chlorophyll.
6. The young leaves of <u>liverworts</u> are known as fiddleheads.
7. In the food-making process, plants <u>take in oxygen and release carbon dioxide</u>.
8. Ferns are <u>nonvascular</u> plants.

Concept Mapping

Complete the following concept map for Section 5–1. Refer to pages B6–B7 to construct a concept map for the entire chapter.

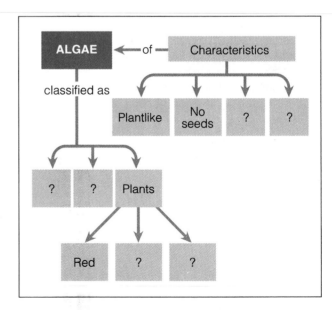

Concept Mastery

Discuss each of the following in a brief paragraph.

1. What are pigments? Why are pigments important to plants?
2. Describe a current or potential use for green algae, red algae, brown algae, mosses, and ferns.
3. How do mosses and their relatives perform the tasks necessary for life on land?
4. How do ferns perform the tasks necessary for life on land?
5. Why can vascular plants grow so much larger than nonvascular plants?
6. What is vascular tissue? Explain why the development of vascular tissue was an important step in the evolution of plants.

Critical Thinking and Problem Solving

Use the skills you have developed in this chapter to answer each of the following.

1. **Making comparisons** How are red, brown, and green algae similar? How are they different?
2. **Applying concepts** A friend tells you that he has seen mosses that were 2 meters tall. Is your friend mistaken? Explain why or why not.
3. **Relating facts** Why do most algae live in shallow water or float on the surface of the water?
4. **Applying concepts** Why can ferns live in drier areas than mosses?
5. **Classifying objects** Examine the accompanying photograph of a plant without seeds. What kind of plant do you think it is? Why? How would you go about confirming your identification?
6. **Using the writing process** Now it's your turn to try being a science teacher. Prepare an outline for a lesson on one of the plants you learned about in this chapter. Your lesson should include an aim, or goal. (For example, the aim of a lesson on classification might be "What are the five kingdoms of living things?") Write a five-question fill-in-the-blank quiz to accompany your lesson and to test whether you have accomplished your aim.

Plants With Seeds

Guide for Reading

After you read the following sections, you will be able to

6–1 Structure of Seed Plants

- Describe the structure of roots, stems, and leaves.
- Give the general equation for photosynthesis.

6–2 Reproduction in Seed Plants

- Discuss the events leading up to the formation of a seed.

6–3 Gymnosperms and Angiosperms

- Describe the four phyla of gymnosperms.
- Describe the structure of a typical flower.

6–4 Patterns of Growth

- Classify plants according to how long it takes them to produce flowers and how long they live.
- Describe some basic ways in which plants grow in response to their environment.

Spotted, striped, or solid-colored; double or single; magenta, pink, scarlet, orange, peach, or white—the brightly colored flowers of the impatiens plant are a familiar sight in suburban gardens and in city window boxes and planters.

Look carefully at the impatiens flowers in the photograph. Can you see a tiny green structure in the center of each flower? After the flower fades and its petals fall off, this structure begins to lengthen, swell, and change color from dark green to a pale yellowish green. Eventually, it is round with a tapered tip and has grooves that run along its length.

Touching the fully grown structure has some startling results. The structure pops off the stem, bursts, and sprays tiny brown objects in all directions. These brown objects are the impatiens's seeds.

Impatiens are just one kind of plant with seeds. What are some other plants with seeds? What do they look like? Where are they found? And why are seeds considered such an important development in plant evolution? Read on to find the answers to these and other questions.

Journal *Activity*

You and Your World   Do you remember when you first learned that plants grow from seeds? When you were very young, did you ever plant seeds and watch them grow? In your journal, write about one of your earliest experiences with seeds.

◄ *The tiny green structure in the center of impatiens flowers develops into a seed pod. When ripe, the seed pod bursts open and scatters the seeds within it.*

6–1 Structure of Seed Plants

Seed plants are among the most numerous plants on Earth. They are also the plants with which most people are familiar. The tomatoes and watermelons in a garden, the pine and oak trees in a forest, and the cotton and wheat plants in a field are all examples of seed plants. What other seed plants can you name?

Seed plants are vascular plants that produce seeds. Recall from Chapter 5 that vascular plants have vascular tissue. Vascular tissue forms a system of tiny tubes that transport water, food, and other materials throughout the body of a plant. There are two types of vascular tissue: **xylem** (ZIGH-luhm) and **phloem** (FLOH-ehm). Xylem carries water and minerals from the roots up through the plant. Because xylem cells have thick cell walls, they also help to support the plant. Phloem carries food throughout the plant. Unlike xylem cells, which carry water and minerals upward only, phloem cells carry materials both upward and downward.

Like all vascular plants, seed plants have true **roots, stems,** and **leaves.** It might be helpful for you to review the adaptations that plants need to survive on land, which are found on pages 115–116. Keep these basic adaptations in mind as you read about roots, stems, and leaves in seed plants.

Figure 6–1 *Plants with seeds come in a wide variety of shapes and sizes. Grasses, orchids, and Douglas firs are all plants with seeds.*

Roots

Roots anchor a plant in the ground and absorb water and minerals from the soil. Roots also store food for plants.

The root systems of plants follow two basic plans: fibrous roots and taproots. Fibrous roots consist of several main roots that branch repeatedly to form a tangled mass of thin roots. Grass, corn, and most trees have fibrous root systems. Taproot systems consist of a long, thick main root (the taproot) and thin, branching roots that extend out of the taproot. Carrots, cacti, and dandelions are examples of plants with taproots.

Have you ever pulled up weeds? If so, you might have noticed that weeds with fibrous roots, such as crabgrass, tend to take a big chunk of soil with them when they are pulled up. Weeds with taproots, such as dandelions, tend to be difficult to pull out of the ground. How does the structure of root systems cause these two problems in weeding?

Refer to Figure 6–2 on page 132 as you read about the structure of a typical root. The outermost layer of the root is called the epidermis (ehp-ih-DER-mihs). The term epidermis (*epi-* means upon; *dermis* means skin) is used to refer to the outermost cell layer of just about any multicellular living thing, including plants, worms, fishes, and humans.

The outer surfaces of the cells of a root's epidermis have many thin, hairlike extensions. These extensions are known as root hairs. The root hairs greatly increase the surface area through which the plant takes in water and minerals from the soil. The water and minerals that are taken up by the root hairs pass into the next layer of the root, the cortex. In many plants, the cells of the cortex store food. They also carry water and dissolved minerals into the center of the root, which is made of vascular tissue, that is, cells of xylem and phloem.

As you can see in Figure 6–2, the very tip of the root is covered by a structure called the root cap. The root cap protects the tip of the root as it grows through the soil. Just behind the root cap is a region that contains growth tissue. This is where new cells are formed.

ACTIVITY
DISCOVERING

Fit to Be Dyed

1. Fill a medium-sized jar one-fourth full of water. Then add a few drops of food coloring and stir. **Note:** *Be careful when using dyes, as they can stain.*

2. Place a stalk of celery in the jar so that its leaves are at the top and its base is at the bottom. Only the base should be submerged in the colored water.

3. Place the jar where it will not be disturbed. After 24 hours, examine your stalk of celery. What do you observe?

4. Remove the stalk of celery from the jar. Using a knife, cut off the portion of the stalk that was under water. Be very careful when using a knife. Discard this portion.

5. Examine the base of the stalk. What do you observe? Why does this occur?

■ Florists sometimes sell red, white, and blue carnations for the Fourth of July and other celebrations. Carnations do not naturally occur in this color combination. How do florists color the carnations without painting them?

■ Obtain a white carnation and test your hypothesis.

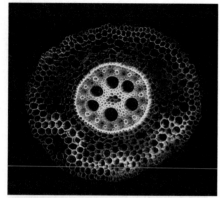

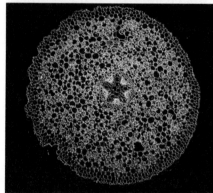

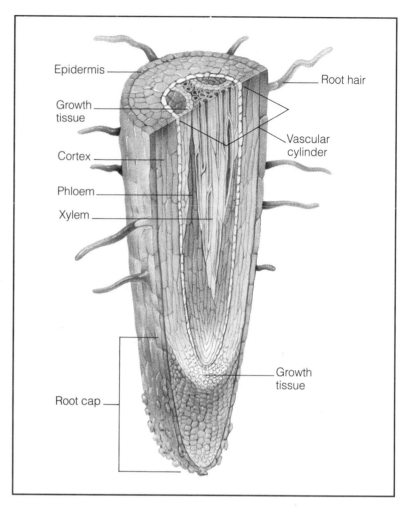

Figure 6–2 *In the root of a buttercup, phloem is located between the arms of a xylem "star" (center left). In the root of a corn plant, yellow circles of xylem alternate with reddish bundles of phloem (top left). Root hairs make this radish sprout look fuzzy (bottom left). What are the functions of the major root structures shown in the diagram?*

For thousands of years, people have used roots in many different ways. Some roots are used for food. Carrots, beets, yams, and turnips are among the many roots that are eaten. The root of the cassava plant is used to make tapioca, which may be familiar to you as the small, starchy lumps in certain kinds of puddings and baby food. Marshmallows were originally candies made from the root of the marsh mallow plant. (Modern marshmallows are made out of sugar, cornstarch, and gelatin.) Roasted chicory and dandelion roots are used as substitutes for coffee. Some roots—licorice, horseradish, and sassafras, for example—are used as spices. Roots are also used to make substances other than food, such as medicines, dyes, and insecticides.

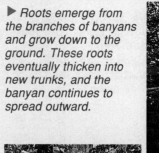

▶ Roots emerge from the branches of banyans and grow down to the ground. These roots eventually thicken into new trunks, and the banyan continues to spread outward.

▲ Thick, sturdy roots that spread out on the surface of the ground form a stable base for certain giant tropical trees.

◀ Tiny roots that emerge from the stems of philodendrons help these plants to climb walls, tree trunks, and other supporting structures.

▲ The mangrove's spreading roots, which look rather like stilts, trap dead leaves and other debris and help create more soil for the plant.

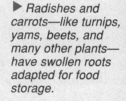

▶ Radishes and carrots—like turnips, yams, beets, and many other plants—have swollen roots adapted for food storage.

Figure 6–3 *Specialized roots*

Stems

Stems provide the means by which water, minerals, and food are transported between the roots and the leaves of a plant. Stems are able to perform this function because they contain xylem and phloem tissue. Stems also hold the leaves of a plant up in the air, thus enabling the leaves to receive sunlight and make food.

Plant stems vary greatly in size and shape. The trunk, branches, and twigs of a tree are all stems. Some plants, such as the strange-looking baobab (BAY-oh-bab) in Figure 6–4 on page 134, have enormous stems that are many meters tall. As you can see, the trunk of the baobab is the most noticeable part of the plant. Other plants, such as the cabbage,

Figure 6–4 *Huge, fat stems and stubby branches make baobab trees look like creatures from another planet. The rectangular patches on the left-hand baobab are places where its bark has been harvested for rope-making.*

Figure 6–5 *Plants bring much color to the world. In spring and summer, flowers such as lupines brighten fields and gardens (bottom). In autumn, the leaves of plants such as the maple turn red and orange (top). Which of these plants is woody? Which is herbaceous?*

have very short stems. The stem of a cabbage is a tough, cone-shaped core that is hidden beneath tightly closed leaves.

Plants may be classified into two groups based on the structure of their stems: herbaceous (her-BAY-shuhs) and woody. Herbaceous plants have stems that are green and soft. Sunflowers, peas, dandelions, grass, and tomatoes are examples of herbaceous plants. As you might expect, woody plants have stems that contain wood. Wood, as defined by plant biologists, is a hard substance made of the layers of xylem that form when a stem grows thicker. Unlike herbaceous stems, woody stems are rigid and quite strong. Roses, maples, and firs are woody plants. What other woody plants can you name?

The structure of a woody stem is shown in Figure 6–6. The outermost layer of the stem is the bark. The outer bark, which is tough and waterproof, helps to protect the fragile tissues beneath it. The innermost part of the bark is the phloem. What is the function of phloem?

The next layer of the stem is called the vascular cambium. The vascular cambium is a growth region of a stem, for it is here that xylem and phloem are produced. The center of the stem is called the pith. It contains large, thin-walled cells that store water and food.

If you were to cut through the stem of certain woody plants, a pattern of rings-within-rings that looks somewhat like a target might be visible. See Figure 6–7. These tree rings are made of xylem.

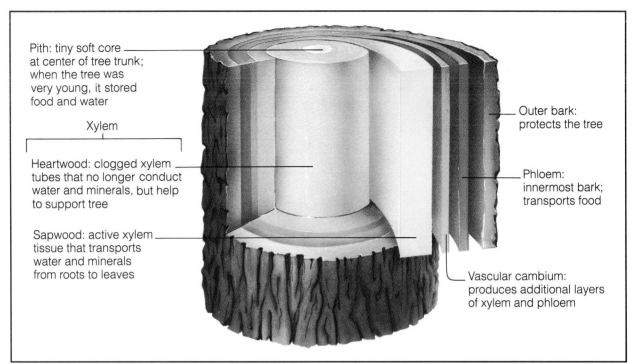

Pith: tiny soft core at center of tree trunk; when the tree was very young, it stored food and water

Xylem

Heartwood: clogged xylem tubes that no longer conduct water and minerals, but help to support tree

Sapwood: active xylem tissue that transports water and minerals from roots to leaves

Outer bark: protects the tree

Phloem: innermost bark; transports food

Vascular cambium: produces additional layers of xylem and phloem

A tree develops rings only if it grows at different rates during the different seasons of the year. For example, many trees in the northern part of the United States grow very little (if at all) in autumn and winter, when the weather is quite cold. They grow rapidly in spring, when it is warm and rainy. And they grow slowly in summer, when it is hot and dry. As a tree grows, a new layer of xylem forms around the old layers. What layer in the stem produces the new xylem cells?

The xylem cells formed in the spring are large and have thin walls. They produce a light brown ring. The xylem cells formed in the summer are small and have thick walls. They produce a dark brown ring. Each pair of light and dark rings represents a year in the life of a tree. How old is a tree with 14 light rings and 14 dark rings?

One of the most important and widely used products of plant stems is wood. Wood is used to

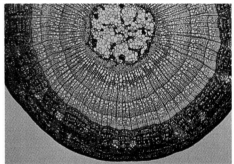

Figure 6–6 *The diagram shows the layers in the stem of a typical tree. The young basswood stem clearly shows its structure. Where is the pith? The vascular cambium? How old is the basswood stem?*

Figure 6–7 *Tree rings can tell you more than just the age of a tree. The spacing and thickness of the rings provide information about weather conditions in the area over time. Thick rings that are far apart indicate years in which conditions were favorable for tree growth—years of much rain. What was the weather like for the first seven years of this larch tree's life? For the last seven years?*

▲ Long stems that run along the surface of the ground allow morning glories to spread into new areas.

▼ Tubers, bulbs, and rhizomes are different types of underground food-storing stems. A potato is a tuber. The "eyes" of the potato are buds. Each bud is capable of developing into a new potato plant. Onions and garlic are bulbs. Bulbs have thick, short leaves that contain food for the plant. Rhizomes such as those of irises grow horizontally along the ground. As they grow, the rhizomes produce buds that grow into new plants.

▲ The thick green stem of a cactus stores water.

Figure 6–8 *Specialized stems*

make a variety of objects ranging from toothpicks to buildings. Ground-up wood is used to make paper. Can you think of some other uses of wood?

Some stems—such as potatoes, onions, ginger, and sugar cane—are a source of food. Other stems are used to make dyes and medicines. Still others have more uncommon uses. For example, the long, flexible stems of certain trees and vines are woven together to make wicker furniture and baskets. And fibers from the stems of flax plants are used to make linen fabric.

Leaves

Plant leaves vary greatly in shape and size. For example, birch trees have oval leaves with jagged edges. Pines, firs, and balsams have needle-shaped leaves. And maples and oaks have flat, wide leaves. Most leaves have a stalk and a blade. The stalk connects the leaf to the stem; the blade is the thin, flat part of the leaf. The thin, flat shape of most leaves exposes a large amount of surface area to sunlight.

Leaves are classified as either simple or compound, depending on the structure of the blade. A simple leaf has a blade that is in one piece. Maple, oak, and apple trees have simple leaves. A compound leaf has a blade that is divided into a number of separate, leaflike parts. Roses, clover, and palms have compound leaves. Can you name some plants that have simple leaves? Compound leaves?

No matter what their shape, leaves are important structures. For it is in the leaves that the sun's energy is captured and used to produce food.

Up to this point, we have been referring to the process in which light energy is used to make food simply as the food-making process. But this important process has its own special name: **photosynthesis** (foht-oh-SIHN-thuh-sihs). The word photosynthesis comes from the root words *photo*, which means light,

ACTIVITY DOING

Dead Ringer

Find a sawed-off tree stump. (Alternatively, find a large circular knot on a piece of wood paneling or piece of lumber.) How old was the tree (or branch) when it was cut? How can you tell? What do the rings tell you about the weather conditions during the life of the tree (or branch)?

Activity Bank

The Ins and Outs of Photosynthesis, p.181

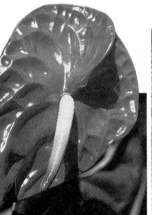

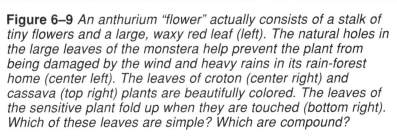

Figure 6–9 *An anthurium "flower" actually consists of a stalk of tiny flowers and a large, waxy red leaf (left). The natural holes in the large leaves of the monstera help prevent the plant from being damaged by the wind and heavy rains in its rain-forest home (center left). The leaves of croton (center right) and cassava (top right) plants are beautifully colored. The leaves of the sensitive plant fold up when they are touched (bottom right). Which of these leaves are simple? Which are compound?*

Figure 6–10 *Plants use the energy of sunlight to make food. The process that uses sunlight, water, and carbon dioxide to make food and oxygen is called photosynthesis.*

ACTIVITY

CALCULATING

It Starts to Add Up

Use a calculator to answer the following questions about an apple tree that has 200,000 leaves, each of which has a surface area of 20 cm².

1. What is the total leaf surface area of the tree?

2. The lower surface of each leaf contains 25,000 stomata per cm². How many stomata are on the lower surface of a single leaf? Assuming that the stomata are found only on the lower surface of the leaves, how many stomata are on the tree?

3. In a single summer, each leaf loses about 86 mL of water through transpiration. How much water is transpired by the entire tree?

and *synthesis*, which means to put together. Can you see why this name is appropriate?

Photosynthesis is the process in which food is synthesized using light energy. Photosynthesis is the largest and most important manufacturing process in the world. It is also one of the most complex. Let's take a closer look at this amazing process.

In photosynthesis, the sun's light energy is captured by chlorophyll, which is the green pigment you read about in Chapter 5. Through a complex series of chemical reactions (which will not be discussed here), the light energy is used to combine water from the soil with carbon dioxide from the air. One of the products of the chemical reactions is food, which is generally in the form of a sugar called glucose.

Glucose can be broken down to release energy. Cells need energy to carry out their life functions. Glucose can also be changed into other chemicals. Some of these chemicals are used by a plant for growth and for repair of its parts. Other chemicals are stored in special areas in the roots and stems.

The other product of photosynthesis is oxygen. Oxygen is important to you, and to almost every other living thing on Earth. Do you know why?

An equation can be used to sum up what occurs during photosynthesis. (Do not be alarmed by the appearance of equations! Equations are simply scientific shorthand for describing chemical reactions.) Here is the equation for photosynthesis in words and in chemical notation:

$$\text{carbon dioxide} + \text{water} \xrightarrow[\text{chlorophyll}]{\text{sunlight}} \text{glucose} + \text{oxygen}$$

$$6\ CO_2 + 6\ H_2O \xrightarrow[\text{chlorophyll}]{\text{sunlight}} C_6H_{12}O_6 + 6\ O_2$$

Why does most photosynthesis occur in leaves? To answer this question, you need to know something about the internal structure of a leaf. Refer to Figure 6–11 as you read about the structure and function of a typical leaf.

The outermost layer of a leaf is called the epidermis. The cells of the epidermis are covered with a

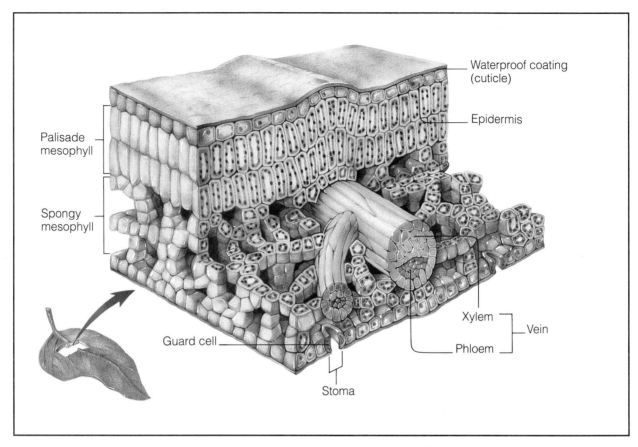

Figure 6–11 *A leaf is a well-designed factory for photosynthesis. How does the structure of each leaf part help the leaf to perform its function?*

Waterproof coating (cuticle)

Epidermis

Palisade mesophyll

Spongy mesophyll

Guard cell

Xylem

Vein

Phloem

Stoma

waxy, waterproof coating that helps to prevent excess water loss. This coating makes some real plants look as if they are made of shiny plastic.

Light passes through the epidermis to reach inner cells known collectively as mesophyll. (*Meso-* means middle; *-phyll* means leaf.) The cells of the mesophyll are where almost all photosynthesis occurs. The shape and arrangement of the upper cells maximize the amount of photosynthesis that takes place. The many air spaces between the cells of the lower layer allow carbon dioxide, oxygen, and water vapor (water in the form of a gas) to flow freely.

Carbon dioxide enters the air spaces within a leaf through microscopic openings in the epidermis. These openings are called stomata (STOH-mah-tah; singular: stoma). The Greek word *stoma* means mouth—and stomata do indeed look like tiny mouths!

In addition to permitting carbon dioxide to enter a leaf, stomata allow oxygen and water to exit. The process in which water is lost through a plant's

ctivity Bank

Bubbling Leaves, p. 183

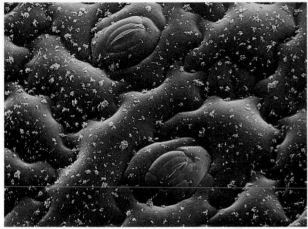
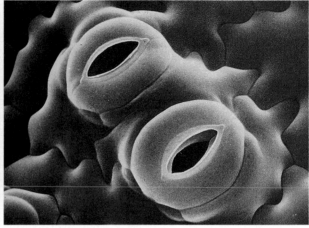

Figure 6–12 *Stomata open to allow carbon dioxide, a raw material of photosynthesis, into the leaf and allow oxygen, a waste product of photosynthesis, out of the leaf. Why do stomata close?*

ACTIVITY

DISCOVERING

Plant-Part Party

Working with a friend or two, make a list of all the foods in your house that come from plants. Remember to include things like ketchup, canned foods, snack foods, and spices. Next to each entry on your list, write the part of the plant from which the food is obtained. If you and your friends cannot identify the source of a food, try to discover its identity using an encyclopedia, cookbook, or botany (plant science) textbook. Compare your list with your classmates'.

■ What foods come from plants?

■ Which plant part was the source of the most foods?

■ One common vegetable looks like stems but is actually the stalks of leaves. What is the name of this vegetable? Were you fooled by these disguised leaves?

leaves is known as transpiration. Plants lose huge amounts of water through transpiration. For example, a full-grown birch tree releases about 17,000 liters of water through transpiration during a single summer season. And the grass on a football field loses almost 3000 liters of water on a summer day.

Although transpiration is necessary for the movement of water through most plants, too much water loss can cause cells to shrivel up and die. One way that plants avoid losing too much water through transpiration is by closing the stomata. Each stoma is formed by two slightly curved epidermal cells called guard cells. When the guard cells swell up with water, they curve away from each other. This opens the stoma. When the water pressure in the guard cells decreases, the guard cells straighten out and come together. This closes the stoma. In general, stomata are open during the day and closed at night. Can you explain why this is so? (*Hint:* When does photosynthesis take place?)

As the cartoon character Popeye tells it, eating spinach is good for you. Popeye is right. In fact, several kinds of leaves are eaten by humans. Spinach, parsley, sage, rosemary, and thyme are just a few examples. What other kinds of leaves are used as food or to season food?

Leaves are also the source of drugs such as digitalis, atropine, and cocaine, of deadly poisons such as strychnine and nicotine, and of dyes such as indigo and henna.

Pitcher plants (left), Venus' flytraps (center), and sundews (right) have leaves that trap insects and other small animals.

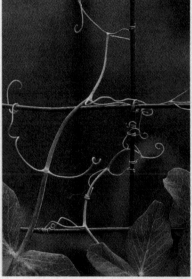

▼ The leaves of a kalanchoe are thick and fleshy, allowing the plant to store lots of water. Kalanchoes also produce tiny offspring plants along the edges of their leaves. Eventually, the leaves fall off and the young plants take root.

▲ The spines of a cactus are actually leaves, as are the threadlike tendrils of peas.

Figure 6–13 *Specialized leaves*

6–1 Section Review

1. What are seed plants?
2. Describe the structure and importance of roots, stems, and leaves.
3. What is photosynthesis? What is the chemical equation for photosynthesis?

Critical Thinking—*Applying Concepts*
4. Stomata are usually located on the lower surface of leaves. Why do plants with floating leaves, such as water lilies, have stomata on the top surface of their leaves?

CONNECTIONS

Plant Power for Power Plants

Imagine a busy power plant: Workers hurry about, checking gauges and adjusting machinery. The air vibrates with the low roar of the giant wheels that spin to produce electricity. Now imagine a peaceful countryside: A tree lifts its branches to the sun. The wind whispers through the leaves. A bird sings. What do these two scenes have in common?

Not that much—right now. However, power plants and green plants may soon use the same chemical reactions to obtain the *fuel* and food they need to function. Scientists recently reported that they are on the brink of producing artificial versions of the "machinery" of photosynthesis.

The reactions of photosynthesis in plants take place in complex, precisely arranged groups of molecules. Special molecules absorb the energy in light and pass it from one set of molecules to the next. The energy runs the chemical reactions that produce food for the plants— that is, glucose.

By copying the machinery of photosynthesis, scientists will be able to produce fuel using some of the least expensive and most abundant raw materials—sunlight, water, and carbon dioxide. With some slight adjustments, hydrogen gas and methane (natural gas), rather than glucose, will be the end products.

Creating artificial forms of photosynthesis may solve the *energy crisis*. And it may also solve two other serious problems facing us—*global warming* and *air pollution*. Here is why. Global warming is caused by excess carbon dioxide in the air. Photosynthesis—and artificial photosynthesis—uses carbon dioxide. So these processes remove excess carbon dioxide from the air, thus reducing global warming. Much air pollution is caused by burning fuels such as oil and gasoline. Hydrogen, on the other hand, does not pollute when it burns. It simply produces water. As you can see, it makes a lot of sense to use plant power in power plants!

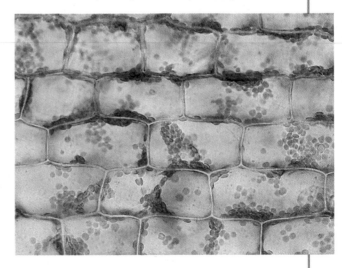

6–2 Reproduction in Seed Plants

Guide for Reading

Focus on this question as you read.

▶ *How is reproduction in seed plants adapted to life on land?*

Seed plants have made two evolutionary leaps that make them fully adapted for life on land. **Seed plants do not require water to reproduce.** Sperm cells are carried straight to the waiting egg; the sperm cells do not have to swim to the egg. **Young seed plants are encased in a structure that provides food and protection.**

Fertilization in Seed Plants

The reproductive structures of seed plants are known as **cones** and **flowers.** Female cones and flower parts contain structures called **ovules** (AH-vyoolz). Each mature ovule contains an egg cell. Male cones and flower parts produce tiny grains of **pollen** (PAHL-uhn), which you can think of as containing sperm cells. The contents of a pollen grain are enclosed in a tough cell wall. This cell wall may be covered with strange and beautiful patterns of spikes and ridges.

If you have allergies, you may know pollen as an irritating form of dust. However, pollen is nothing to sneeze at! Carried by the wind or by animals, each grain of pollen is capable of delivering a sperm cell to an egg cell. The process by which pollen is carried from male reproductive structures to female reproductive structures is called **pollination.**

If everything goes right, pollination is followed by fertilization. The contents of the pollen grain break

Figure 6–14 *A breeze causes pollen from a cluster of male pine cones to drift away in a dusty yellowish cloud. A few of these pollen grains may eventually reach a female pine cone and fertilize its ovules. Some seed plants, such as the African tulip tree, have flowers rather than cones.*

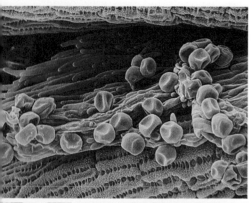

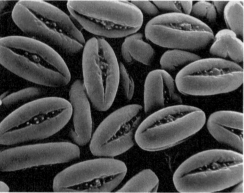

Figure 6–15 *Pollen grains may be spherical or oval, textured or smooth. The pollen grains from a flowering horse chestnut resemble loaves of French bread.*

out of their hard cell wall. They then grow a long tube that delivers the sperm cell to the egg cell in the ovule. The sperm cell joins with the egg cell, fertilizing it. The fertilized egg and the ovule that surrounds it develop into a seed.

In some seed plants, the ovules and the seeds that develop from them are contained within a structure called the ovary (OH-vah-ree). These plants are known as **angiosperms** (AN-jee-oh-sperms). The Greek word *angion* means vessel; *sperma* means seed. Thus angiosperms are plants whose seeds are contained in a vessel (the ovary).

In other seed plants, the ovules and seeds are not surrounded by an ovary. These plants are known as **gymnosperms** (JIHM-noh-sperms). The Greek word *gymnos* means naked. Thus, gymnosperms are plants whose seeds are naked; that is, not covered by an ovary.

Seeds

Although seeds look quite different from one another, they all have basically the same structure. **A seed consists of a seed coat, a young plant, and stored food.**

The seed coat is a tough, protective covering that develops from the ovule wall. Some familiar seed coats include the "skins" on lima beans, peanuts, and corn kernels. The brown, winglike covering on pine seeds is also a seed coat. So is the fleshy red covering of a yew seed.

Figure 6–16 *Fertilized ovules develop into seeds. The seeds of gymnosperms, such as pines, are not covered by an ovary. Some of the seeds in this pine cone have been removed from their normal position on the top of the woody scales that make up the cone so that they can be clearly seen (top left). The seeds of angiosperms—such as cantaloupes (top right), pomegranates (bottom left), and avocados (bottom right)—are enclosed by an ovary.*

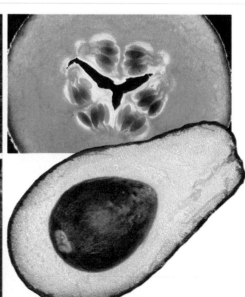

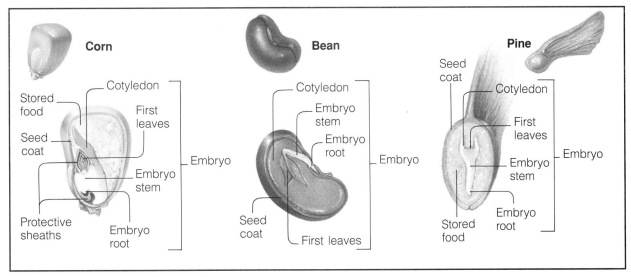

Corn

- Stored food
- Cotyledon
- First leaves
- Seed coat
- Embryo
- Embryo stem
- Protective sheaths
- Embryo root

Bean

- Cotyledon
- Embryo stem
- Embryo root
- Embryo
- Seed coat
- First leaves

Pine

- Seed coat
- Cotyledon
- First leaves
- Embryo
- Embryo stem
- Stored food
- Embryo root

Enclosed within the seed coat is a tiny young plant, or embryo. The embryo develops from the fertilized egg. As you can see in Figure 6–17, the embryo is basically a miniature plant. The top of the embryo eventually gives rise to the leaves and upper stem of the plant. The middle stemlike portion of the embryo becomes the bottom part of the plant's stem. The bottom portion of the embryo becomes the plant's roots.

You probably have noticed that the embryo makes up only a small portion of the seed. In the seeds of most plants, the embryo stops growing when it is quite small and enters a state of "suspended animation." While "sleeping" inside the seed, the embryo can survive long periods of cold, heat, or dryness. When conditions are favorable for growth, the embryo becomes active and the young plant begins to grow once more. Once the young plant begins to grow, it uses the food stored in the seed until it can make its own.

Figure 6–17 In some seeds, such as those of pines, the embryo is surrounded by stored food. In other seeds, such as corn kernels, some of the food is stored inside a seed leaf, or cotyledon. In still other seeds, such as beans, all of the food is stored inside the cotyledons. Why do most seeds consist mostly of stored food?

Figure 6–18 These tropical plants have seeds and fruits that are quite exotic in appearance. Can you identify their seeds and fruits?

ACTIVITY
DISCOVERING

Seed Germination

1. Obtain 20 dried beans or unpopped popcorn kernels, 20 test tubes, and some paper towels.

2. Design a two-part experiment that uses these materials to determine whether light and warmth are needed for the seeds to germinate.

3. Write down what you plan to do in your experiment and what results you expect.

4. Soak the seeds in water overnight. Then put together the apparatus for your experiment.

■ What were the results of your experiment? Did the results match your predictions? What were some possible sources of error?

Seed Dispersal

After seeds have finished forming, they are usually scattered far from where they were produced. The scattering of seeds is called seed dispersal (dih-SPER-suhl). Seeds are dispersed in many ways. Have you ever picked a dandelion puff and blown away all its tiny fluffy seeds? If so, you have helped to scatter the seeds. Usually, dandelion seeds are scattered by the wind. Other plants whose seeds are scattered by the wind include maples and certain pines. The winglike structures on these seeds cause them to spin through the air like tiny propellers.

Humans and animals play a part in seed dispersal. For example, burdock seeds have spines that stick to people's clothing or to animal fur. People or animals may pick up the seeds on a walk through a field or forest. At some other place, the seeds may fall off and eventually start a new plant.

The seeds of most water plants are scattered by floating in oceans, rivers, and streams. The coconut, which is the seed of the coconut palm, floats in water. This seed is carried from one piece of land to another by ocean currents.

Other seeds are scattered by a kind of natural explosion, which sends the seeds flying into the air. This is how the impatiens you read about at the beginning of this chapter disperses its seeds.

Figure 6–19 *The seeds of the coconut palm are dispersed by water (center). A tumbleweed is said to disperse its seeds mechanically. The main body of the plant serves as a device for scattering seeds (top right). The seeds of the unicorn plant hitch a ride to new places when spines hook around the hooves (or hiking shoes) of animals that walk by (bottom right). How do you think milkweed seeds are dispersed (left)?*

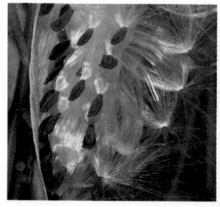

Most seeds remain dormant, or inactive, for a time after being scattered. If there is enough moisture and oxygen, and if the temperature is just right, most seeds go through a process called germination (jer-mih-NAY-shuhn). Germination is the early growth stage of an embryo plant. Some people call this stage "sprouting."

6–2 Section Review

1. Explain how reproduction in seed plants is adapted to life on land.
2. What is a seed? What are the main parts of a seed?
3. How do angiosperms differ from gymnosperms?

Critical Thinking—*Sequencing Events*
4. Put these events in the correct order: germination, fertilization, seed dispersal, release of pollen, seed formation, pollination. Briefly describe each event.

PROBLEM ??? Solving

They Went Thataway!

Seeds are dispersed in many different ways. Seed dispersal is important because it helps plants spread to new areas. It also improves the chances that some of a plant's seeds will grow and survive to produce seeds of their own. Examine the accompanying photographs carefully. Then answer the following questions for each photograph.

Making Inferences
1. How is the seed dispersed? How can you tell?
2. How is the seed or fruit adapted for this method of dispersal?

Figure 6–20 *Cycads (top) and ginkgoes (bottom) are gymnosperms that were quite common during the age of the dinosaurs. Unlike a conifer, a cycad or a ginkgo produces either male cones or female cones, not both.*

6–3 Gymnosperms and Angiosperms

Gymnosperms are the most ancient group of seed plants. They first appeared about 360 million years ago—about the same time as the first land animals. Throughout the age of the dinosaurs, approximately 65 to 245 million years ago, gymnosperms were the dominant form of plant life on Earth.

Near the end of the age of dinosaurs, about 100 million years ago, a new phylum of plants appeared on the scene. The plants in this new phylum—the angiosperms—soon replaced the gymnosperm phyla as the Earth's dominant form of plant life.

Gymnosperms

Although the reign of gymnosperms has ended, four phyla of gymnosperms have survived to the present day. **The four phyla of gymnosperms are commonly known as cycads, ginkgoes, conifers, and gnetophytes.**

Cycads (SIGH-kadz) are tropical plants that look like palm trees. Some cycads grow up to 15 meters tall, but most grow no taller than a human. The trunk of a cycad is topped by a cluster of feathery leaves. In the center of the leaves of a mature cycad are its cones.

Although ginkgoes (GING-kohz) were fairly common during the age of the dinosaurs, only one species exists today. The ginkgo is sometimes known as the maidenhair tree. Interestingly, the ginkgo does not seem to exist in the wild. The ginkgoes that grow along many city streets in the United States and elsewhere are the descendants of plants from gardens in China. Ginkgo seeds have an incredibly bad odor.

With about 550 species, conifers (KAHN-ih-ferz) make up the largest group of gymnosperms. Most conifers are large trees with needlelike or scalelike leaves. See Figure 6–21. Conifers are found throughout the world, and are the dominant plants in many forests in the Northern Hemisphere. Conifers include the world's tallest trees (coast redwoods) and longest-lived trees (bristlecone pines).

The word conifer literally means cone-bearer. This name is quite appropriate. Like cycads and ginkgoes, most conifers bear cones that produce pollen or ovules.

A few conifers, such as larches and bald cypresses, shed their leaves in autumn. However, most conifers—such as pines, firs, spruces, cedars, and hemlocks—are evergreen. Evergreen plants have leaves year round. Old leaves drop off gradually and are replaced by new ones throughout the life of the plant. The needles or leaves may remain on an evergreen tree for 2 to 12 years.

Conifers are of great importance to people. They are a major source of wood for building and for manufacturing paper. Useful substances such as turpentine, pitch, and rosin are made from the sap of conifers. The seeds of certain pines are rich in protein and are used for cooking and snacking. Juniper seeds are used to flavor food. Recently, scientists discovered that taxol, a substance in the bark of the Pacific yew tree, is showing promise as an effective treatment for certain kinds of cancer.

Gnetophytes (NEE-toh-fights) are a diverse group of plants that share characteristics with both gymnosperms and angiosperms. Gnetophytes include climbing tropical vines with oval leaves, bushes with jointed branches and tiny scalelike leaves, and a peculiar-looking desert plant whose two straplike leaves grow throughout its lifetime.

Figure 6–21 *Some conifers, such as the Port Oxford cedar, have tiny scalelike leaves (left). Others, such as the Douglas fir, have needlelike leaves (center). A few conifers are amazingly large. Why is this giant redwood tree able to live and grow in spite of having a car-sized tunnel cut through it?*

Figure 6–22 Welwitschia mirabilis *is one of the few organisms that can survive in the extremely hot, dry Namib desert of Africa. Although* Welwitschia *has only two leaves, the leaves soon become tattered and torn so that they look like many leaves. To which phylum of gymnosperms does* Welwitschia *belong?*

Figure 6–23 *Some flowers, such as sweet alyssum, impatiens, geraniums, and marigolds, are common garden flowers (top left). Others, such as the strange silversword plant of Hawaii, are delightfully uncommon (right). The silversword looks quite different after it develops its enormous stalk of flowers (bottom left). After it produces its seeds, the silversword dies.*

Angiosperms

Angiosperms make up the largest group of plants in the world—there are more than 230,000 known species, with perhaps hundreds of thousands more as yet undiscovered.

Angiosperms vary greatly in size and shape. They can be found in just about all of Earth's environments, including frozen wastelands near the North Pole, steamy tropical jungles, and practically waterless deserts.

Because they produce flowers, angiosperms are also called flowering plants. The flowers of angiosperms fill the Earth with beautiful colors and pleasant smells. But flowers serve a more important purpose. **Flowers are the structures that contain the reproductive organs of angiosperms.**

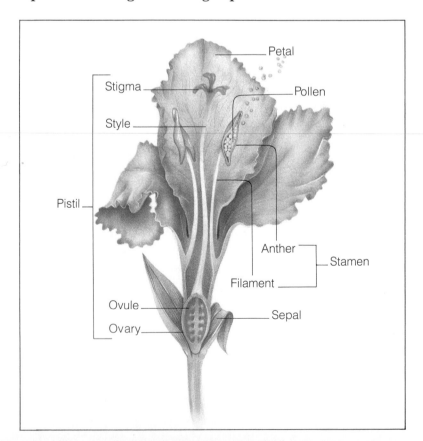

Figure 6–24 *The parts of a flower work together to accomplish the task of reproduction. Which are the male parts of a flower? Which are the female parts?*

Figure 6–25 *Flowers and the animals that pollinate them have evolved in response to one another. How are the hummingbird, bee, and bat adapted to feeding from flowers? Can you tell what kind of animal pollinates the bee orchid, which looks and smells like a female bumblebee (top right)? Plants have also evolved ways that help them to disperse their seeds. How are blackberries adapted for seed dispersal by animals?*

As you can see in Figure 6–23, flowers come in all sorts of sizes, colors, and forms. As you read about the parts of a typical flower, keep in mind that the descriptions do not apply to all flowers. For example, some flowers have only male reproductive parts, and some flowers lack petals.

When a flower is still a bud, it is enclosed by leaflike structures called **sepals** (SEE-puhls). Sepals protect the developing flower. Once the sepals fold back and the flower opens, colorful leaflike structures called **petals** are revealed. The colors, shapes, and odors of the petals attract insects and other animals. These creatures play a vital role in the reproduction of flowering plants.

Within the petals are the flower's reproductive organs. The thin stalks topped by small knobs are the male reproductive organs, or **stamens** (STAY-muhns). The stalklike part of the stamen is called the filament, and the knoblike part is called the anther. The anther produces pollen.

The female reproductive organs, or **pistils** (PIHS-tihls), are found in the center of the flower. Some

ACTIVITY

READING

The Birds and the Bees

The process of evolution has resulted in some amazing relationships between angiosperms and the animals that pollinate them. Discover some of these fascinating relationships in *The Clover and the Bee: A Book of Pollination* by Anne Dowden.

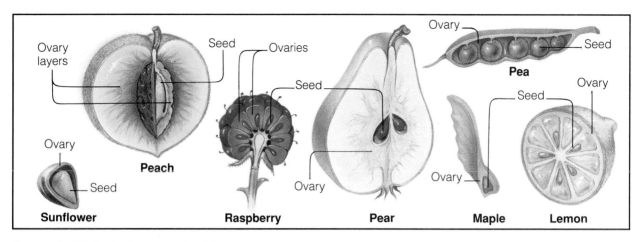

Ovary layers — Seed — Peach

Ovary — Seed — Sunflower

Ovaries — Seed — Raspberry

Seed — Ovary — Pear

Ovary — Seed — Pea

Seed — Ovary — Maple

Seed — Ovary — Lemon

Figure 6–26 *Fruits have evolved in ways that help to disperse seeds. How are maple seeds specialized for being dispersed by the wind? How are peaches, pears, lemons, and raspberries adapted for seed dispersal by animals?*

A World of Fun With Plants

Do you know how to make a shrill whistle with a blade of grass? Learn how to do this and other neat things by reading *Hidden Stories in Plants* by Anne Pellowski.

flowers have two or more pistils; others have only one. The sticky tip of the pistil is called the stigma. A slender tube, called the style, connects the stigma to a hollow structure at the base of the flower. This hollow structure is the ovary, which contains one or more ovules.

A flower is pollinated when a grain of pollen lands on the stigma. (Can you explain why the stigma is sticky?) If the pollen is from the right kind of plant, the wall of the pollen grain breaks open. The contents of the pollen grain then produce a tube that grows down through the style and into an ovule. When the tube has finished growing, a sperm cell emerges from the tube and fertilizes the egg cell in the ovule.

Like gymnosperms, some angiosperms are pollinated by the wind. The flowers of these plants are usually small, unscented, and produce huge amounts of pollen. Most angiosperms, however, are pollinated by insects, birds, and other animals. Flowers that rely on animals for pollination are colorful, scented, and/or full of food. This helps them to attract the animals that pollinate them.

As you have learned, after the egg cell is fertilized, the ovule develops into a seed. As the seed develops, the ovary also undergoes some changes and becomes a fruit. A **fruit** is a ripened ovary that encloses and protects the seed or seeds. Apples and cherries are fruits. So are many of the plant foods you usually think of as vegetables, such as cucumbers and tomatoes. What other kinds of fruits do you eat? The next time you are enjoying a favorite fruit, try to identify the seed, seed coat, and ripened ovary.

6–3 Section Review

1. What are the four phyla of gymnosperms?
2. What is a flower? How are flowers involved in reproduction in angiosperms?
3. Describe pollination and fertilization in angiosperms.

Connection—*You and Your World*
4. Why are almost all the ginkgoes that are planted male trees?

6–4 Patterns of Growth

Why do some garden plants die and have to be replaced each year? Why do you have to turn a houseplant that is growing on the windowsill every now and then? To answer questions such as these, you have to know something about the patterns of growth in plants.

Annuals, Biennials, and Perennials

Plants are placed into three groups according to how long it takes them to produce flowers and how long they live. The three groups of plants are **annuals, biennials,** and **perennials.**

Some plants grow from a seed, flower, produce seeds, and die all in the course of one growing season. Such plants include marigolds, petunias, and many other common garden flowers. **Plants that complete their life cycle within one growing season are called annuals.** The word annual comes from the Latin word *annus,* which means year. Most annuals have herbaceous (nonwoody) stems. Wheat, rye, and tobacco plants are other examples of annuals.

Plants that complete their life cycle in two years are called biennials. (The Latin word *bi* means two.) Biennials sprout and grow roots, stems, and leaves during their first growing season. Although the stems and leaves may die during the winter, the roots survive. During the second growing season, a

Focus on these questions as you read.
▶ *How do annuals, biennials, and perennials differ from one another?*
▶ *How do plants grow in response to their external environment?*

Figure 6–27 *Petunias bloom cheerfully throughout the warm months of the year, but die in the winter (top). During its second and final growing season, the foxglove produces a spike of many flowers. The spotted "trail" on a foxglove flower directs bees to the nectar and pollen inside (right). Begonias live for many years, producing flowers each summer (bottom). Which of these plants is a perennial? Which is an annual? Which is a biennial?*

biennial grows new stems and leaves and then produces flowers and seeds. Once the flowers produce seeds, the plant dies. Examples of biennials include sugar beets, carrots, celery, and certain kinds of foxgloves.

Still other plants live for more than two growing seasons. **Plants that live for many years are called perennials.** (The Latin word *per* means through; a perennial lives through the years.) Some perennials, such as most garden peonies, are herbaceous. Their leaves and stems die each winter and are renewed each spring. Most perennials, however, are woody. The long lives of perennials permit them to accumulate lots of layers of woody xylem in their stems. Pine trees and rhododendron bushes are examples of woody perennials. What other plants are woody perennials?

Tropisms

When studying behavior in plants and other living things, it is helpful to be familiar with two terms: stimulus (STIHM-yoo-luhs; plural: stimuli, STIHM-yoo-ligh) and response. A stimulus is something in a

ctivity Bank

Lean to the Light, p. 184

living thing's environment (both its internal and external environment) that causes a reaction, or response.

Plants respond to stimuli in a variety of ways. One way is by adjusting the way they grow. The growth of a plant toward or away from a stimulus is called a **tropism** (TROH-pihz-uhm). If a plant grows toward a stimulus, it is said to have a positive tropism. If it grows away from a stimulus, it is said to have a negative tropism.

There are several kinds of tropisms. The two most important tropisms involve responses to light and to gravity.

All plants exhibit a response to light called phototropism. (Recall that the Greek word *photos* means light.) Think about the houseplant mentioned at the start of this section. If a houseplant is not turned regularly, it soon begins to bend toward the window, the source of its light. Do leaves and stems show positive or negative phototropism?

Plants also show a response to gravity, or gravitropism. Roots show positive gravitropism—they grow downward. Stems, on the other hand, show negative gravitropism—they grow upward, against the pull of gravity.

Figure 6–28 *Because this houseplant was not turned regularly, it started to grow in a crooked way. Why did this happen?*

6–4 Section Review

1. What are annuals, biennials, and perennials?
2. What are tropisms? Describe two tropisms.

Critical Thinking—*Making Predictions*
3. Suppose a tree is knocked over by a storm, but it survives the experience. What will this tree look like a few years after the storm?

ACTIVITY
DISCOVERING

Pick a Plant

In the late winter or early spring, purchase a packet of seeds for an annual plant you would like to grow. You may wish to get together with a classmate or two and split the cost of the seeds. (A packet usually costs $1 to $2.) Following the directions on the seed packet, plant 3 seeds in a clean pint-sized milk carton. Water the seeds and resulting plants every day or two.

■ How long does it take your seeds to sprout? How do your plants change over the course of the year? Make a poster showing the more interesting stages in the life cycle of your plant. The stages shown in your poster should include the seed, the plant when it first begins to sprout, the plant when it has two leaves, the plant when it develops flowers, and the plant when it develops fruits or seeds.

Laboratory Investigation

Gravitropism

Problem

How does gravity affect the growth of a seed?

Materials (per group)

4 corn seeds soaked in water for 24 hours	paper towels
	masking tape
petri dish	glass-marking pencil
scissors	clay

Procedure

1. Arrange the four seeds in a petri dish. The pointed ends of all the seeds should face the center of the dish. One of the seeds should be at the 12 o'clock position of the circle, and the other seeds at 3, 6, and 9 o'clock.

2. Place a circle of paper towel over the seeds. Then pack the dish with enough pieces of paper towel so that the seeds will be held firmly in place when the other half of the petri dish is put on.

3. Moisten the paper towels with water. Cover the petri dish and seal the two halves together with a strip of masking tape.

4. With the glass-marking pencil, draw an arrow on the lid pointing toward 12 o'clock. Label the lid with your name and the date.

5. With pieces of clay, prop the dish up so that the arrow is pointing up. Place the dish in a completely dark place.

6. Predict what will happen to the seeds. Then observe the seeds each day for about one week. Make a sketch of them each day. Be sure to return the dish and seeds to their original position when you have finished.

Observations

What happened to the corn seeds? In which direction did the roots and the stems grow?

Analysis and Conclusions

1. Which part of the germinating seeds showed positive gravitropism? Which part showed negative gravitropism?

2. What would happen to the corn seeds if the dish were turned so that the arrow was pointed toward the bottom of the dish? If it were turned to the right or the left?

3. Why is it important that the petri dish remain in a stable position throughout the investigation?

4. Explain why the seeds were placed in the dark rather than near a sunny window.

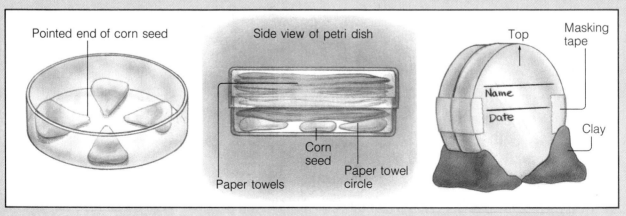

Pointed end of corn seed   Side view of petri dish   Top   Masking tape

Name   Date   Clay

Paper towels   Corn seed   Paper towel circle

Summarizing Key Concepts

6–1 Structure of Seed Plants

▲ There are two types of vascular tissue. Xylem carries water and dissolved minerals. Phloem carries food.

▲ Roots anchor the plant in the ground and absorb water and minerals from the soil.

▲ Stems transport materials between the roots and the leaves and support the leaves.

▲ In most plants, leaves are the structures in which photosynthesis occurs.

▲ Photosynthesis is the process that uses light energy to change carbon dioxide and water into glucose and oxygen.

6–2 Reproduction in Seed Plants

▲ Seed plants do not require standing water to reproduce.

▲ The process in which pollen is carried from male reproductive structures to female reproductive structures is called pollination.

▲ After pollination, the contents of the pollen grain grow a tube into an ovule and release a sperm cell. The sperm cell fuses with the

ovule's egg cell to produce a fertilized egg in a process called fertilization.

▲ The ovules and seeds of angiosperms are covered by an ovary; those of gymnosperms are uncovered.

▲ Seeds consist of a seed coat, an embryo, and stored food.

6–3 Gymnosperms and Angiosperms

▲ There are four living phyla of gymnosperms: cycads, ginkgoes, conifers, and gnetophytes.

▲ Because their reproductive structures are contained in flowers, angiosperms are known as flowering plants.

6–4 Patterns of Growth

▲ Plants are classified as annuals, biennials, and perennials according to how long it takes them to produce flowers and how long they live.

▲ The growth of a plant in response to an environmental stimulus is called a tropism.

Reviewing Key Terms

Define each term in a complete sentence.

6–1 Structure of Seed Plants
xylem
phloem
root
stem
leaf
photosynthesis

6–2 Reproduction in Seed Plants
cone

flower
ovule
pollen
pollination
angiosperm
gymnosperm

6–3 Gymnosperms and Angiosperms
sepal
petal
stamen

pistil
fruit

6–4 Patterns of Growth
annual
biennial
perennial
tropism

Chapter Review

Content Review

Multiple Choice

Choose the letter of the answer that best completes each statement.

1. Roots grow away from light. This is an example of
   a. negative gravitropism.
   b. positive gravitropism.
   c. negative phototropism.
   d. positive phototropism.
2. Which phylum has seeds that are enclosed by an ovary?
   a. gnetophytes   c. conifers
   b. angiosperms   d. cycads
3. The movement of gases in and out of the leaf is regulated by the
   a. stomata.   c. cortex.
   b. mesophyll.   d. cambium.
4. Many flowering plants rely on animal partners for
   a. pollination.   c. fertilization.
   b. germination.   d. transpiration.
5. Photosynthesis produces
   a. chlorophyll.   c. water.
   b. carbon dioxide. d. oxygen.

6. The main water-conducting tissue in a plant is
   a. phloem.   c. cambium.
   b. xylem.   d. pith.
7. The functions of anchoring the plant and absorbing water are performed primarily by a plant's
   a. roots.   c. leaves.
   b. stems.   d. anthers.
8. Which structure is not part of a flower's pistil?
   a. style   c. ovary
   b. stigma   d. anther
9. A pine tree is best described as
   a. annual.   c. biennial.
   b. perennial.   d. herbaceous.
10. The union of the sperm cell and the egg cell is known as
   a. pollination.   c. fertilization.
   b. germination.   d. transpiration.

True or False

If the statement is true, write "true." If it is false, change the underlined word or words to make the statement true.

1. Pollen is produced in the <u>anther</u>.
2. The <u>cortex</u> produces new xylem and phloem cells.
3. A four-leafed clover is an example of a <u>simple</u> leaf.
4. Carrots and dandelions have <u>fibrous root</u> systems.
5. The protective outer covering of a seed is called the <u>epidermis</u>.
6. The ovules of a conifer are located in its <u>flowers</u>.
7. The structures that protect a flower bud are called <u>stamens</u>.
8. A <u>fruit</u> is a ripened ovary.

Concept Mapping

Complete the following concept map for Section 6–1. Refer to pages B6–B7 to construct a concept map for the entire chapter.

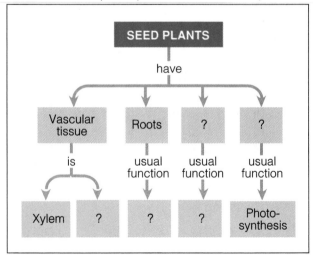

Concept Mastery

Discuss each of the following in a brief paragraph.

1. List the three main parts of a seed and describe their function.
2. Give three examples of ways in which people use conifers.
3. What is seed dispersal? How does it take place? Why is it important?
4. Distinguish between herbaceous and woody plants. Would you expect an annual plant to be woody? Explain.
5. How does the structure of a stem help it to carry out its functions?
6. What is photosynthesis?
7. How are angiosperms and gymnosperms similar? How are they different? Give three examples of each kind of plant.
8. Describe the two most important kinds of tropisms in terms of stimulus and response.
9. Unlike the first plants to appear on Earth, seed plants do not depend on water for reproduction. Explain why.

Critical Thinking and Problem Solving

Use the skills you have developed in this chapter to answer each of the following.

1. **Summarizing information** List the major parts of a typical flower. Briefly describe the function of each part.
2. **Relating concepts** Describe the adaptations in seed plants that help them to avoid excess water loss.
3. **Developing a hypothesis** You observe that hungry deer have eaten the bark on an apple tree as far up as they can reach. At first, the tree seems to remain healthy. However, it eventually dies. Develop a hypothesis to explain your observations. How might you test your hypothesis?
4. **Identifying relationships** When life first appeared on Earth more than 3.5 billion years ago, there was no oxygen gas in the air. But for the past 1.8 billion years or so, the air has been about 20 percent oxygen. How did photosynthesis cause this change in the composition of Earth's air and maintain it? Why is it important to most of Earth's organisms to have so much oxygen in the air?
5. **Making inferences** Pesticides are designed to kill harmful insects. Sometimes, however, they kill helpful insects. What effect could this have on angiosperms?
6. **Designing an experiment** Design an experiment to determine whether or not water is needed for seed germination. Describe your experimental setup. Be sure to include a control.
7. **Using the writing process** Imagine that you are the seed plant of your choice. Describe your life from your earliest memories as a forming seed to the time you produced seeds of your own. (*Hints:* What were seed dispersal and germination like? What changes did you undergo as you grew?)

Colleen Cavanaugh Explores the Underwater World of

TUBE WORMS

The seabed lies 2500 meters below the ocean's surface. Here, the pressure is nearly 260 times that at the Earth's surface, and the temperature is close to the freezing point. The region is always dark, as sunlight cannot penetrate these ocean depths.

Almost all organisms depend on light as their source of energy in manufacturing food. So scientists had expected to find only the simplest creatures inhabiting this dark area of the Pacific Ocean floor near the Galápagos Islands. To the surprise of a group of scientists from the Woods Hole Oceanographic Institution, however, communities of strange sea animals were found in this forbidding environment.

Among the animals observed by the Woods Hole team are giant clams that measure one-third meter in diameter, oversized mussels, and crabs. But perhaps the most striking organisms are giant tube worms. These worms are so different from anything seen before that scientists have placed them in a new family of the animal kingdom.

Although most of the worms are only 2.5 to 5 centimeters in diameter, they can be as long as 2 meters! Like a sausage in a tube, each worm lives inside a tough rigid casing. One end of the tube worm is anchored to the seabed. At the other end, a red plume-like structure made of blood-filled filaments waves in the ocean water.

Internally, the body of the worm is most unusual. Approximately half of its body is made of a colony of densely packed bacteria. These bacteria are the key to a worm's ability to survive in such a harsh environment. But exactly what is the relationship between the bacteria and the tube worms?

Enter Dr. Colleen Cavanaugh, a marine biologist and microbiologist who studies these bacteria and their relationship to the larger life forms they support. She has found

that the bacteria break down hydrogen sulfide, a common sulfur compound, to produce energy. This energy is then used by the tube worms. In this way, the seemingly worthless hydrogen sulfide—which smells like rotten eggs and is poisonous to most organisms—provides the energy for a fascinating aquatic ecosystem.

Hydrogen sulfide is not ordinarily found in sea water. But plenty of hydrogen sulfide is found near hydrothermal vents such as those in the Pacific Ocean areas where the tube worms were first discovered. Hydrothermal vents are openings in the ocean floor that allow ocean water to come into contact with the earth's hot, molten interior. In the vents, ocean water is heated to temperatures as high as 350 degrees Celsius. The hot water is ejected back into the ocean, laced with many different chemical compounds, including hydrogen sulfide. Tube worms and other animals that are able to use sulfur abound in these regions.

Research on tube worms requires interest, dedication, and a broad knowledge of science. So it is not surprising that Colleen Cavanaugh is part of the team at Woods Hole exploring this fascinating underwater world. As a high-school student growing up in Michigan, Colleen Cavanaugh explored many fields of science. She then went on to study general ecology and biology at the University of Michigan. As a graduate student at Harvard University, she developed an interest in microbial ecology. Now she combines these three fields, as well as others such as oceanography, in her work. The work Dr. Cavanaugh has already done on tube worms will be of help in her more general study of the nature of symbiosis, or cooperation, between animals and bacteria. As she points out, the sulfides and sulfide-based ecosystems such as that of the tube worms are not limited just to the ocean depths. They are also found in salt marshes, mud flats, and many other marine environments closer to home. Thus, scientific work in the dark reaches of the sea may shed new light on the nature of life on the surface as well.

▲ Living on the ocean bottom near hydrothermal vents (top), tube worms (center) thrive as a result of a symbiotic relationship with bacteria (bottom).

PESTS OR PESTICIDES:
WHICH WILL IT BE?

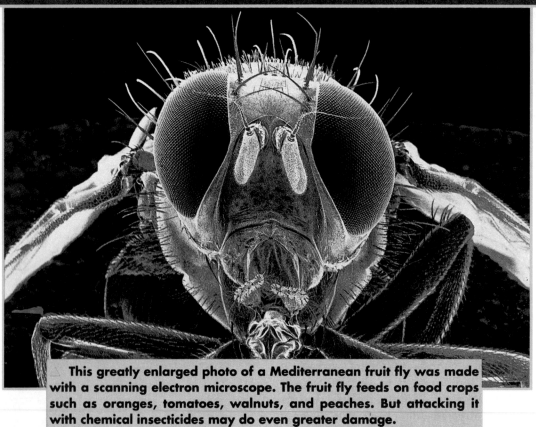

▲ This greatly enlarged photo of a Mediterranean fruit fly was made with a scanning electron microscope. The fruit fly feeds on food crops such as oranges, tomatoes, walnuts, and peaches. But attacking it with chemical insecticides may do even greater damage.

There is a war being fought right now that has been going on for thousands of years. It is a fight for survival against armies that far outnumber the Earth's entire human population. It is the war people wage against pests, such as certain insects, rats, mice, weeds, and fungi.

Some insects, rats, and mice carry deadly diseases such as malaria, bubonic plague, and typhoid fever. Throughout history, diseases carried by pests have killed hundreds

of millions of people. These animals also threaten food supplies around the world.

Other threats to world food supplies are weeds and fungi. Weeds compete with crops for water and nutrients. Some fungi cause plant diseases that result in huge crop losses. If people lose the war against these organisms, human disease, suffering, and starvation will occur worldwide.

Today people have many weapons to use in the fight against pests. During the past few decades, scientists have made or discovered many chemicals that kill pests. These chemicals, which are called pesticides, have helped to increase food production by killing the pests that destroy crops and eat stored food. Pesticides have also saved many human lives by killing disease-carrying pests.

Unfortunately, people have not always used pesticides wisely. Farmers and others who use pesticides have accidentally killed useful organisms such as honeybees, spiders, songbirds, fish, falcons, dogs, and cats. In addition, the overuse of pesticides has made some pests resistant to the pesticides. In other words, the pests have become able to withstand assaults of deadly chemicals and are now much harder to kill.

To make matters worse, pesticides are sometimes washed into rivers and streams, where they kill fish and other animals. Winds can spread pesticides over hundreds or thousands of kilometers. This pollutes areas far from where the chemical was used to kill pests. Pesticides have injured and killed people in all parts of the world.

So now we face a difficult problem. How can we save crops and kill disease-carrying pests without harming the environment?

THE OTHER SIDE OF PESTICIDES

There are now about 35,000 different pesticides on the market. Each year in the Unit-ed States alone, about one-half billion kilograms of these chemicals are used by farmers, homeowners, and industry.

Specific pesticides called herbicides and fungicides have helped farmers to reduce crop losses due to weeds and harmful fungi. Insect-killing pesticides, or insecticides, have not been as successful in reducing losses. In fact, the amount of crops destroyed by harmful insects has nearly doubled during the past 30 years—even though farmers have been using more powerful insecticides in greater quantities than ever before.

Insecticides are also failing to protect people from insect-borne diseases. For example, malaria, a sometimes fatal disease that is carried by certain mosquitoes, has made an alarming comeback in countries where it had been practically wiped out.

Why are the insecticides failing to protect people and crops? One reason is that the use of insecticides has caused harmful insects to become resistant to them. When an area is sprayed with pesticides, most of the pests are killed. A few individuals, however, may be naturally resistant to the pesticide and may survive. These survivors produce offspring that, like themselves, are resistant. In a short time, all of the pests are resistant.

Another reason is that insecticides can also kill the natural enemies of insect pests. For example, insecticides have killed ladybugs

▼ **One way to combat insect pests is to spray them from the air with chemical insecticides. But this method, called crop dusting, also affects living things other than pests.**

Farmers use pesticides against such destructive insects as the tobacco hornworm. The hornworm attacks many kinds of crops, including these tomato plants. However, pesticides can also harm helpful animals and even people.

in some apple orchards. Now there are no more ladybugs in these orchards to keep apple-eating insects, such as aphids, under control.

Because of the harmful effects of pesticides, some people believe that these chemicals are too dangerous to use. But those in favor of using pesticides disagree. They say that pesticides would not be so dangerous if people used them properly.

NEW WEAPONS

One way to lessen our dependence on pesticides is to understand how crops and insects interact with their environments. The study of how living things interact with their environments is called ecology. Ecology provides clues to how to control pests by changing the environment of either the pest or its victim.

For example, a plant called barberry often carries a fungus that causes a disease called black stem rust. When barberry grows near wheat, the rust spreads from the barberry to the wheat. Some farmers have protected their wheat from black stem rust by destroying nearby barberry plants rather than by spraying fungicide.

Other diseases can be controlled by crop rotation. Periodically planting crops other than cabbage in a cabbage field is an example. This method prevents disease-causing organisms that attack only cabbage from building up in the soil.

Pests may also be controlled by finding naturally-occurring things that are harmful only to the pests—preferably things to which the pests cannot become resistant. For example, Japanese beetles are killed by dusting an area with bacteria that causes a fatal beetle disease. Other insects are sprayed with synthetic versions of their own body chemicals. This upsets their normal chemical balance and causes them to grow abnormally and then to die. And because no pest can build up a resistance to being eaten, farmers may introduce helpful insect-eating animals—praying mantises, spiders, and ladybugs, for example—into their fields and orchards.

Not all new types of pest control are so down-to-earth, however. By studying how the crop-eating desert locust interacts with its environment, scientists have been able to use satellites to fight this pest.

The desert locust is a flying insect related to the grasshopper. From time to time, millions of these locusts gather and sweep across Africa and India, eating every crop and blade of grass in their path. No weapons have been able to stop these insects once they take flight. But in recent years, satellites orbiting the Earth have been taking pictures of areas in Africa and India where locusts might breed. Scientists can identify possible breeding areas by the amount of moisture they contain. If satellite photos show that an area is moist enough for locust eggs to survive there, scientists warn the people that may be threatened. The people can then concentrate their pesticide spraying in these areas.

The better we understand how living things interact with their environments, the more clues we will find for controlling pests without using large amounts of chemicals. Perhaps someday we will not have to use chemicals at all!

The Corn Is As High As A Satellite's Eye

Farms of the future will move into space, and crops will be far out, too.

The old man held his granddaughter's hand as the high-speed elevator zoomed 180 floors to the top of the Triple Towers. Out on the observation deck, the young girl stared in wonder at the huge city that fanned out to meet the horizon in all directions. Far below, crowds of little dots moved along in neat columns. "Look at all those people!" the girl exclaimed. "There must be millions of them."

The old man nodded. "Twenty-two million, to be exact," he said softly.

▲ Short food plants, such as lettuce and many other kinds of vegetables, can be grown within revolving drums. On Earth, these devices make it possible for many plants to be grown in a small amount of space. In the future, similar devices may simulate gravity and thereby enable orbiting space stations to grow food crops.

"But Grandpa, how do all these people keep from starving? I read that years ago people in cities all over the world couldn't get enough to eat. And there are even more people now than there were back then."

The old man frowned as he remembered his own youth. "Yes, that's true, Lisa," he said with a sigh. "Twenty percent of the world was going hungry when I was a young man in 1992. The farmers could use only a quarter of the Earth's land for farming."

"Why couldn't they use more of the land, Grandpa?" Lisa asked.

"Well," he answered, "the rest was either sizzling hot deserts, barren mountains, or frozen tundras. Of the available farmlands, some has to be used for nonfood crops like cotton. Less food was available and more and more people were being born."

"Are people still starving, Grandpa?" Lisa asked in a worried voice.

"Well, dear, some are, but because of techniques that were developed during the late twentieth century, farmers can now produce more food."

"Oh, Grandpa, that sounds interesting. Will you tell me how more food is made?"

"Of course I will, Lisa. Let's sit down on that bench and I'll tell you all about it," the old man said with a smile.

SPACE CROPS

"Look up there. What do you see?" the old man asked.

"Stars, Grandpa. Why?"

"Well, Lisa, much of the world's food is now being grown in space. We have space stations that contain rotating drums filled with plants. The rotation simulates the Earth's gravity. The plants are bathed in fluorescent light, which simulates the sun. Crops that do not grow very tall or that grow in the ground, like lettuce, potatoes, radishes, and carrots, are grown in the drums," the old man stated proudly.

"Wow, that's neat, Grandpa. I never knew that salad came from outer space," Lisa said excitedly.

Laughing, the old man said, "Well, most of it does. We also have space stations where plants grow out of the sides of walls or are moved along conveyor belts through humid air."

GENETICS MAKES THE DIFFERENCE

"Is all our food grown in outer space, Grandpa?" asked Lisa, curious.

"Why, of course not. Methods of growing plants on Earth have been improved too. When I was young, Lisa, 20 percent of the crops that were grown on Earth died off. Now, because of changes scientists have made in the genetic material of plants, very few crops die. The genetic changes have resulted in plants with increased resistance to temperature changes, diseases, herbicides, and harsh environments. Plants have been made resistant by several methods. Some plants were altered by a process called protoplast fusion," the old man said.

"That sounds complicated."

"It's not really very complicated," he replied. "The plant cell walls are dissolved by enzymes. The part remaining is called a protoplast. This protoplast is then fused to another protoplast from a different strain of the same plant or from an entirely different plant. The cell wall is regrown around the fused protoplasts, and a new plant develops. We now have many varieties of such plants. And you might think it strange, but many are smaller than those that used to be grown."

"Smaller? But that doesn't make sense," Lisa protested.

"Well, the smaller plants can produce as much as or more than the big ones because they have been genetically changed to be highly productive," explained the old man. "Also, more of them can be grown in the same amount of space. Two examples of the smaller plants are dwarf rice plants and dwarf peach trees."

"Wow, that's neat. Tell me more," Lisa said.

"All right. We also have plants that can get the nitrogen they need directly from the air. Plants need nitrogen to build amino acids, proteins, and other cell chemicals. In the old days, most plants got their nitrogen from the soil. But this process was using up all the nitrogen in the soil, and the plants didn't grow well. So fertilizer had to be added to the soil. Soybeans, legumes, alfalfa, and clover have always been able to use nitrogen directly from the air with the help of nitrogen-fixing bacteria. Other plants that can use nitrogen directly from the air were developed by combining genes from these nitrogen-fixing bacteria with genes from plants that normally took their nitrogen from the soil. Now almost all the plants take their

▲ Dwarf trees being grown right now may one day supply enough fruit to fill most of our needs. What is the advantage of growing dwarf plants?

nitrogen directly from the endless supply in the air."

"I didn't know that. Are any other plants different from the plants you saw when you were young?" Lisa asked.

"Oh, yes," the old man declared. "Many of the crops are now considered to be high-yield. That means they can be harvested faster than they could before, so that more can be grown. By redesigning the plants genetically, scientists have shortened the plants' growing time."

"It seems that genetic changes have kept people from starving," Lisa said thoughtfully. "That and the growing of food in space or on land."

"That's true," said her grandfather. "But there's still a third place. Some of the crops are grown in the ocean. This leaves more room for crops that can be grown only on land."

"That's a neat story, Grandpa!"

"It's not a story, Lisa. It's the truth. But scientists still have to keep working to improve crop production because the population continues to grow," the old man observed.

For Further Reading

If you have been intrigued by the concepts examined in this textbook, you may also be interested in the ways fellow thinkers—novelists, poets, essayists, as well as scientists—have imaginatively explored the same ideas.

Chapter 1: Classification of Living Things

Jarrell, Randall. *The Animal Family.* New York: Pantheon.

Paterson, Katherine. *Jacob Have I Loved.* New York: Avon.

Chapter 2: Viruses and Monerans

Christopher, John. *No Blade of Grass.* New York: Simon & Schuster.

De Kruif, Paul. *Microbe Hunters.* New York: Harcourt, Brace & World, Inc.

Chapter 3: Protists

Foster, Alan Dean. *The Thing.* New York: Bantam.

Perez, Norah A. *The Passage.* Philadelphia: Lippincott.

Chapter 4: Fungi

Cameron, Eleanor. *The Wonderful Flight to the Mushroom Planet.* Boston: Little Brown and Co.

Hughes, William Howard. *Alexander Fleming and Penicillin.* Hove, U.K.: Wayland. LTD.

Jones, Judith, and Evan Jones. *Knead It, Punch It, Bake It! Make Your Own Bread.* New York: Thomas Y. Crowell.

Chapter 5: Plants Without Seeds

Abels, Harriette E. *Future Food.* Mankato, MN: Crestwood House.

Kavaler, Lucy. *The Wonders of Algae.* New York: The John Day Co.

Sterling, Dorothy. *The Story of Mosses, Ferns, and Mushrooms.* Garden City, NY: Doubleday & Co.

Chapter 6: Plants With Seeds

Busch, Phyllis B. *Wildflowers and the Stories Behind Their Names.* New York: Charles Scribner's Sons.

Elbert, Virginie Fowler. *Grow a Plant Pet.* Garden City, NY: Doubleday & Co.

Activity Bank

Welcome to the Activity Bank! This is an exciting and enjoyable part of your science textbook. By using the Activity Bank you will have the chance to make a variety of interesting and different observations about science. The best thing about the Activity Bank is that you and your classmates will become the detectives, and as with any investigation you will have to sort through information to find the truth. There will be many twists and turns along the way, some surprises and disappointments too. So always remember to keep an open mind, ask lots of questions, and have fun learning about science.

A KEY TO THE PUZZLE

One way of showing how objects are classified is a taxonomic key. A taxonomic key consists of many pairs of opposing descriptions. Only one of the descriptions in a pair is correct for a given object. Following the correct description is an instruction that directs you to another pair of descriptions. By following each successive description and instruction in a taxonomic key, you will eventually arrive at an object's correct classification group.

Don't worry—this sounds more complicated than it actually is. Let's take a look at a simple taxonomic key for the five kingdoms.

1a Made up of cells that contain a nucleus. Go to 2.

1b Made up of one cell that does not contain a nucleus. *Monera.*

2a Is unicellular. Go to 3.

2b Is multicellular. Go to 4.

3a Belongs to the green, red, or brown algae phylum. *Plantae.*

3b Does not belong to the green, red, or brown algae phylum. Go to 7.

4a Is an autotroph. *Plantae.*

4b Is a heterotroph. Go to 5.

5a Evolved from autotrophs. *Plantae.*

5b Did not evolve from autotrophs. Go to 6.

6a Has a cell wall. *Fungi.*

6b Does not have a cell wall. *Animalia.*

7a Is a yeast. *Fungi.*

7b Is not a yeast. *Protista.*

1. Using the taxonomic key for the five kingdoms, determine the kingdom to which each of the organisms described in the following stories belongs.

a. While looking at a sample of pond water through a microscope, you notice a rod-shaped unicellular heterotroph that has a cell wall and no nucleus.

b. In the same sample, you see a unicellular heterotroph that has a nucleus and moves by means of whiplike "tails." Because yeasts don't have "tails," you know it is not a yeast.

c. While walking through the woods one day, you notice some white, candy-cane-shaped multicellular heterotrophs growing on a dead log. The cells contain a nucleus and have a cell wall. Chemical study of the hereditary material reveals that they are descended from autotrophs.

d. In a tide pool, you see an autotroph that looks like sheets of green cellophane. It is made up of many cells that contain a nucleus. It looks exactly like the photograph in your field guide of the green alga known as sea lettuce.

e. Further examination of the organisms in the tide pool reveals a small, nonmoving, multicellular blob. At first, you think it's some kind of seaweed, but you later find out that it is a heterotroph whose cells lack cell walls.

2. According to the taxonomic key, how are animals and fungi similar? Different?

3. Invent a taxonomic key for pets. Test your key by using it to determine the kind of pet a friend or classmate is thinking of.

A GERM THAT INFECTS GERMS

In everyday speech, the term germ refers to any microorganism that causes disease. Because bacteriophages infect bacteria, you can say that they are germs that infect germs! In this activity you will build a model of a bacteriophage so that you can better understand how it works.

1. Use the diagram on page B40 and any of the following materials to build your own model bacteriophage: pipe cleaners, construction paper, screws, nuts, bolts, scissors, tape, crayons, glue, screwdriver.

2. Make a drawing of your model and label the parts. Which part contains the hereditary material? Which part would the bacteriophage use to attach itself to a bacterial cell?

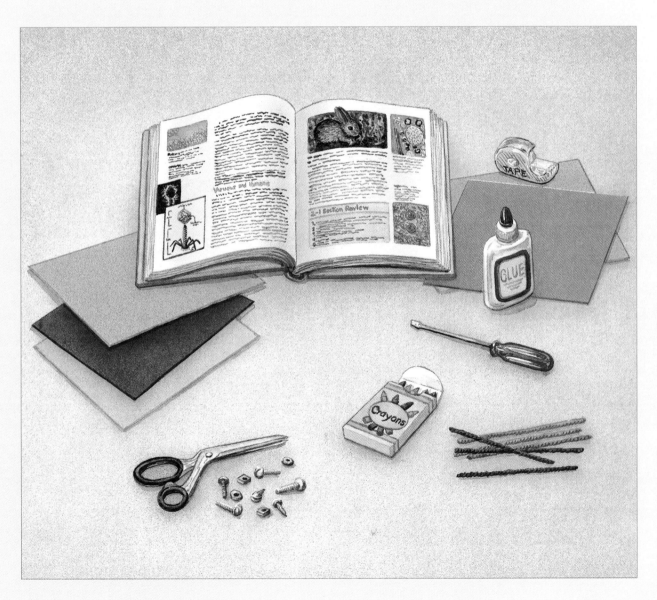

YUCK! WHAT ARE THOSE BACTERIA DOING IN MY YOGURT?

Introduction

Even newly opened, fresh-from-the-refrigerator yogurt is loaded with bacteria. But don't throw your yogurt away in disgust. Bacteria are supposed to be in yogurt. Why? Find out in this activity.

Materials

50 mL milk
10 mL diluted
    chocolate syrup
10 mL lemon juice
10 mL tea
10 mL vinegar

10 mL water
pH paper
5 paper cups
graduated cylinder
5 spoons

If you don't have a graduated cylinder, use a measuring spoon. One teaspoon equals 5 mL.

Procedure

1. Assign roles to each member of the group. Possible roles include: Recorder (the person who records observations and coordinates the group's presentation of results), Materials Manager (the person who makes sure that the group has all the materials it needs), Maintenance Director (the person who coordinates cleanup), Principal Investigator (the person who reads instructions to the group, makes sure that the proper procedure is being followed, and asks questions of the teacher on behalf of the group), and Specialists (people who perform specific tasks such as preparing the cups of liquids to be tested or testing the pH of the liquids). Your group may divide up the tasks differently, and individuals may have more than one role, depending on the size of your group.

2. On a separate sheet of paper, prepare a data table similar to the one shown in the Observations section.

3. Label the paper cups Chocolate, Lemon, Tea, Vinegar, and Water. Pour the appropriate liquid into each cup.

4. Dip a pH strip into each liquid. Compare the color of the strip to the key on the pH paper box. Record the pH in the appropriate place in your data table.

5. Pour 10 mL of milk into each of the five cups. Use a different spoon to mix the milk with each of the other liquids. Record your observations in your data table.

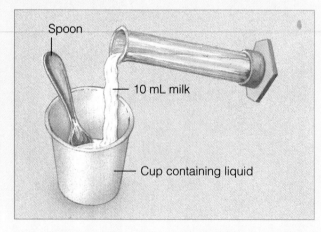

6. Measure the pH of the mixture in each cup. Record the pH in the appropriate place in your data table.

Observations

DATA TABLE

Liquid	pH Before	Observations	pH After
Chocolate			
Lemon			
Tea			
Vinegar			
Water			

Analysis and Conclusions

1. Describe what happened to the milk when it was added to the vinegar. Look at the substances formed when the milk reacted with the vinegar. How might these substances relate to the production of yogurt?

2. Did any other liquids cause the milk to change in the same way as the vinegar?

3. Relate the pH of the liquid to the way it reacts with milk.

4. Using what you have observed in this activity, explain why yogurt is thick and almost solid whereas milk is a thin liquid.

5. The essential ingredients of yogurt are milk and bacteria. (The bacteria are called "active cultures" on the ingredients label on a container of yogurt.) Without the bacteria, most yogurt would simply be fruit-flavored milk. What is the purpose of the bacteria in yogurt?

PUTTING THE SQUEEZE ON

Amebas are not the only protists with contractile vacuoles. In fact, most freshwater protists have contractile vacuoles. In this activity you will observe how this tiny structure works under different conditions.

Materials

3 *Paramecium caudatum* cultures at room temperature (25°C): fresh water, 0.5 percent table salt solution, 1.0 percent table salt solution

refrigerated (2°C) *Paramecium caudatum* culture in fresh water

medicine dropper

4 glass microscope slides

4 coverslips

microscope

cotton ball

Procedure 🧪

1. Using a medicine dropper, put a drop of water containing *Paramecium caudatum* in the center of a clean microscope slide.

2. Pull apart a small piece of cotton and put a few threads in the drop of water.

3. Cover the drop with a coverslip.

4. Locate and focus on a paramecium. Notice the alternating contractions of the two contractile vacuoles. How long does it take for a contractile vacuole to contract, refill, and contract once again? Record this information in the appropriate place in a data table similar to the one shown in Observations.

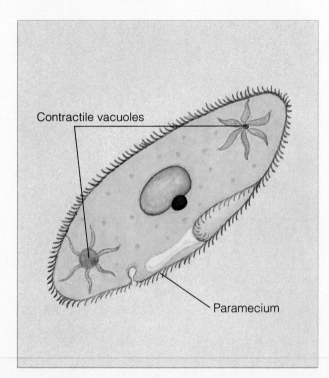

Contractile vacuoles

Paramecium

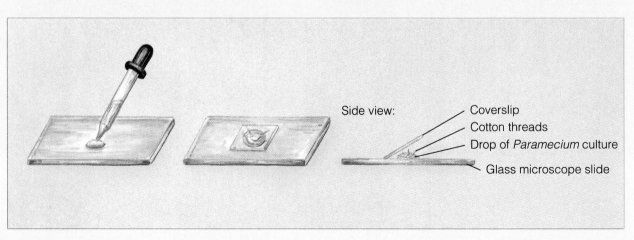

Side view:
Coverslip
Cotton threads
Drop of *Paramecium* culture
Glass microscope slide

5. Repeat steps 1 through 4 for a paramecium in a 0.5 percent table salt solution, a 1.0 percent table salt solution, and very cold water. Record your data.

6. Observe a paramecium in the very cold water. What happens as the water warms up?

Observations

DATA TABLE

Percent Table Salt	Temperature	Time Between Contractions
0	25°C	
0.5	25°C	
1	25°C	
0	2°C	

Analysis and Conclusions

1. What is the purpose of contractile vacuoles?

2. How does an increase in the concentration of salt in its environment affect a paramecium's contractile vacuoles?

3. What can you infer about the rate at which water enters the paramecium in salt solutions? Explain.

4. How is the rate at which a paramecium performs life functions, such as pumping out excess water, affected by very cold temperatures? What evidence do you have for this?

5. Compare your results with those obtained by your classmates. Did you obtain the same results? Why or why not?

SHEDDING A LITTLE LIGHT ON EUGLENA

The plantlike protists known as euglenas are quite popular in scientific research—they are small, relatively easy to take care of, and fairly simple in structure. In Section 3-3, you read about about the structure of euglenas. In this activity you will discover how the euglenas' structure enables them to function in a particular situation.

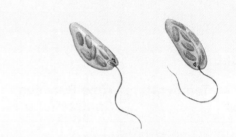

Procedure and Observations

1. Pour a concentrated culture of *Euglena* into a petri dish. What color is the concentrated culture? Why is the culture this color?

2. Cover half the dish with aluminum foil or a small piece of cardboard.

3. After 10 minutes, uncover the dish. What do you observe?

Analysis and Conclusions

1. How do euglenas respond to light?

2. Why do you think this occurs?

3. What structures in euglenas make this possible?

Have you ever opened a bag of bread and found fuzzy black mold growing on the slices you had planned to use? Or discovered spots of mildew that seemed to appear overnight in a clean bathroom? You may wonder how fungi manage to get just about everyplace. In this activity you will discover the secret of fungi's success.

What Do I Do?

1. Obtain a round balloon, cotton balls, tape, a stick or ruler about 30 cm long, modeling clay, and a pin.

2. Stretch the balloon so that it inflates easily. Do not tie off the end of the balloon!

3. Pull a cotton ball into five pieces about the same size. Roll the pieces into little balls no more than 1 cm in diameter.

4. Insert the little cotton balls through the opening in the neck of the balloon.

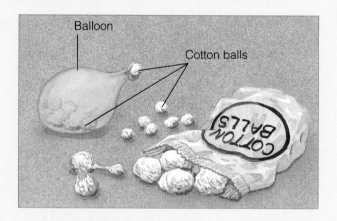

5. Continue making little cotton balls and putting them into the balloon until the balloon is almost full.

6. Inflate the balloon and tie a knot in its neck to keep the air inside.

7. Tape the knotted end of the balloon to the top of the stick.

8. Put the bottom of the stick into the modeling clay. Shape the modeling clay so that the stick stands upright.

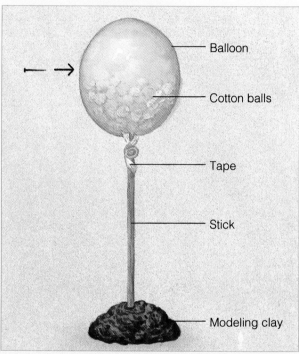

9. You have now made a model of the fruiting body of a bread mold. The balloon represents the spore case, the cotton balls represent spores, and the stick represents the stalk (a hypha) that supports the spore case. Make a drawing of your model. Label the spores, spore case, and stalk.

10. After you have finished admiring your model, jab the balloon with the pin.

What Did I Learn?

1. What happened when you jabbed the balloon with the pin?

2. Relate what you observed to reproduction in fungi. Why are fungi found just about everywhere?

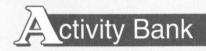
YEAST MEETS BEST

What do yeasts need to grow? What are the best conditions for growing yeast? Find out by doing this activity.

What You Need

5 small narrow-necked bottles (the 355-mL or 473-mL plastic or glass ones used for soda and juice work well)

china marker or marking pen

5 round balloons, stretched so that they inflate easily

5 plastic straws

sugar

salt

warm (40-45°C) water

activated yeast

250 mL beaker or small glass

graduated cylinder or measuring spoons (1 teaspoon equals 5 mL)

What To Do

1. Using the china marker, label the bottles A, B, C, D, and E.
2. Fill the bottles about two-thirds full with warm water.
3. Put 5 mL of sugar into bottle A.
4. Put 30 mL of sugar into bottles B, C, and E.
5. Put 5 mL of salt into bottles C, D, and E.
6. Put 2 mL of dry powdered yeast into bottle A and stir with a clean straw. Remove the straw.
7. Quickly place a balloon over the opening of bottle A. Make sure that the balloon fits tightly around the neck of the bottle.

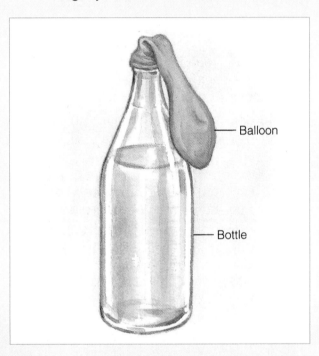

Balloon

Bottle

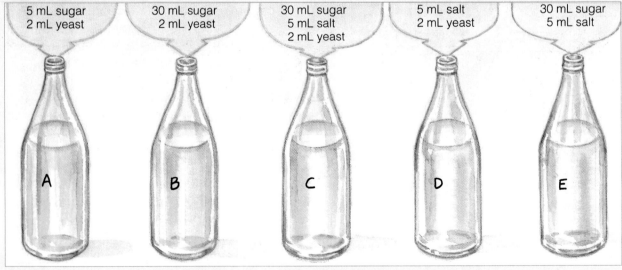

A	B	C	D	E
5 mL sugar 2 mL yeast	30 mL sugar 2 mL yeast	30 mL sugar 5 mL salt 2 mL yeast	5 mL salt 2 mL yeast	30 mL sugar 5 mL salt

8. Repeat steps 6 and 7 for bottles B, C, and D.

9. Repeat step 7 for bottle E.

10. Place the bottles in a warm spot away from drafts. Observe them carefully. Record your observations in a data table similar to the following.

What Did You Observe?

DATA TABLE

	Sugar	Salt	Yeast	Observations
A	5 mL	none	2 mL	
B	30 mL	none	2 mL	
C	30 mL	5 mL	2 mL	
D	none	5 mL	2 mL	
E	30 mL	5 mL	none	

What Did You Learn?

1. What happened to the balloons on some of the bottles? Why do you think this happened?

2. Relate what you observed to the growth of yeast.

3. Relate what you observed to how yeasts make the bubbles in bread, beer, and champagne.

4. What seem to be the best conditions for growing yeast?

5. Design an experiment for determining how temperature affects the growth of yeast.

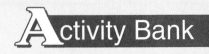
SEAWEED SWEETS

How do people use seaweed? You can experience how seaweed is used as a food by using this old family recipe to make a traditional Chinese dessert. The name of this dessert is based on the Chinese dialect called Cantonese. *Popo* means maternal grandmother. *Kanten* means agar in general and this dessert in particular. Kanten is sold in oriental grocery stores. It comes in long, thin blocks.

Popo Thom's Kanten

2 blocks of kanten (one 0.5 oz package)

1 cup sugar

4 cups liquid (Popo Thom uses guava juice. You can use any kind of fruit juice you like.)

food coloring (optional)

Recipe Directions

1. Rinse kanten under running water.
2. Break kanten into cubes and put the cubes into a saucepan.
3. Add the liquid and sugar to the saucepan. Bring the mixture to a boil.

Kanten
1 cup sugar dissolved in 4 cups liquid

Then turn down the heat and cook, stirring, until the kanten dissolves.

4. Remove the saucepan from the heat. If you wish, stir in a few drops of food coloring to make the mixture a pretty color.
5. Pour the mixture into a shallow rectangular pan.
6. Let cool; then refrigerate until set.
7. Cut kanten into blocks and serve. Do you like the dessert, or do you think it's an acquired taste?

Another Chinese dessert, almond float, is made in a similar way. In step 3, use 2 cups of water and 2 tablespoons of almond extract as the liquid. In step 4, stir in 2 cups of evaporated milk. Almond float cubes are usually served with canned fruits in their syrup (canned lichee, or litchi nuts, are traditional; fruit cocktail is also good).

Things to Think About

1. The ingredients in a recipe often serve a particular function. For example, cornstarch thickens sauces, mustard helps the oil and vinegar in a salad dressing to mix, and egg helps to "glue" other ingredients together. What purpose does the kanten serve in this recipe? What purposes do the sugar and juice serve?
2. Share the kanten you have made with your family and/or classmates. What do they think of this dessert?
3. Working with some friends, brainstorm some other ways in which you might use kanten blocks.

THE INS AND OUTS OF PHOTOSYNTHESIS

Introduction

Blue-green bacteria, plantlike protists, and green plants perform a special process known as photosynthesis (foh-toh-SIHN-thuh-sihs). In photosynthesis, light energy is used to combine carbon dioxide and water to form food and oxygen. As you read on pages B111–112 in the textbook, oxygen is very important to humans and many other living things. Oxygen, which we humans take in when we breathe in, is used to break down food to release energy—energy that powers all life processes. The breakdown of food also produces carbon dioxide, which is removed from the body when we breathe out. What do organisms that can perform photosynthesis do with carbon dioxide?

In this activity you will observe the "ins and outs" of photosynthesis and the complementary process of food breakdown, or respiration.

Materials

3 125-mL flasks
3 #5 rubber stoppers
100 mL graduated cylinder
bromthymol blue solution
2 sprigs of *Elodea*, about the same size
light source
drinking straw

Procedure 🧪 🧰

1. Using the graduated cylinder, measure out 100 mL of bromthymol blue solution for each of the three flasks. **CAUTION**: *Bromthymol blue is a dye and can stain your hands and clothing*. What color is the bromthymol blue solution?

2. Put the straw into one of the flasks. Gently blow bubbles into the solution until there is a change in its appearance. How does the solution change? Repeat this procedure with the other two flasks.

3. Place one sprig of *Elodea* in each of two of the flasks. Stopper all three flasks.

4. Put one flask containing *Elodea* in the dark for 24 hours. Put the other two flasks on a sunny windowsill for the same amount of time.

5. After 24 hours, examine each flask. What do you observe?

(continued)

On a Sunny Windowsill

Stopper

Elodea

In the Dark

Analysis and Conclusions

1. What substance did you add to the bromthymol blue solution when you blew bubbles into it? Where did this substance come from? What effect did this substance have on the solution?

2. What happened in the flask that was kept in the dark? Explain why this occurred.

3. What happened in the flask containing *Elodea* that was kept in a sunny spot? Why?

4. What was the purpose of the flask that did not contain *Elodea*?

5. How are the "ins," or raw materials, for photosynthesis related to the "outs," or products, of respiration?

6. How are the "outs" of photosynthesis related to the "ins" of respiration?

7. How is photosynthesis related to respiration?

Photosynthesis uses the gas carbon dioxide and produces the gas oxygen. Tiny openings in the leaves called stomata allow carbon dioxide to get in and oxygen to get out. Where are stomata located? Are they in the same position in all leaves? Discover for yourself by doing this activity.

Procedure and Observations

1. Obtain a variety of fresh leaves and a glass container of hot water. **CAUTION:** *Be careful when handling hot substances.*

2. Pinch the stalk of a leaf. Then dip the flat part of the leaf in hot water. What do you observe? Why do you think this happens?

3. Repeat step 2 with some different kinds of leaves. How do your results differ from one another?

Analysis and Conclusions

1. What can you infer about the structure of the leaves?

2. Share your findings with your classmates. Compare your results. Are they similar? Why or why not? What are some possible sources of error in this activity?

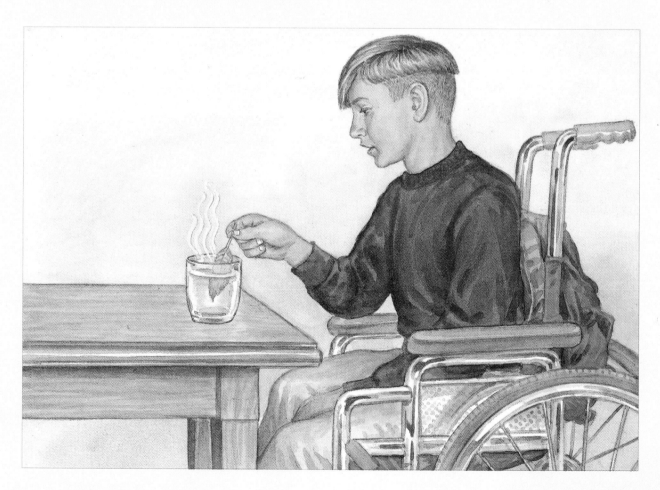

LEAN TO THE LIGHT

Living things respond to their environment. Pet dogs and cats come running when they hear the can opener or the rustle of the pet-food bag. People put up umbrellas and hurry toward shelter when it starts to rain. In this activity you will discover one way in which plants respond to their environment.

What Do I Need?

2 half-gallon milk cartons with the tops cut off, old newspapers, old artist's smock or apron, black tempera (poster) paint, paintbrush, 2 plastic cups, soil, scissors, and 10 popcorn kernels.

What Do I Do?

1. Spread some old newspapers on a work surface. Wear old clothes on which it is okay to get paint, or wear an artist's smock or apron to protect your clothes. Paint the inside of the milk cartons black. Put the cartons on a sheet of old newspaper someplace where they can dry, and then clean up your work area.

2. Fill the cups about three-fourths full with soil. Water the soil so that it is moist but not soaking wet. Plant five popcorn kernels in each cup. The kernels should be just under the surface of the soil.

3. Cut a small hole about 1 cm across in the side of one of the dry milk cartons.

4. Set the cups in a well-lighted place. Cover each cup with a milk carton. Make sure to position the carton with the hole in it so that light can enter through the hole.

5. Examine your seedlings after several days. If the soil is dry, add a little water. Replace the cartons so that they are in exactly the same position they were originally.

What Did I Find Out?

1. Describe how your seedlings appear after they have been growing for a week. How are the two groups of seedlings different?

2. Why do you think this happens?

3. Why might your results be important to someone who has houseplants?

4. Draw a picture that show your results. Using the picture, share your findings with your classmates.

Appendix A

The metric system of measurement is used by scientists throughout the world. It is based on units of ten. Each unit is ten times larger or ten times smaller than the next unit. The most commonly used units of the metric system are given below. After you have finished reading about the metric system, try to put it to use. How tall are you in metrics? What is your mass? What is your normal body temperature in degrees Celsius?

Commonly Used Metric Units

Length   The distance from one point to another

meter (m)   A meter is slightly longer than a yard.
1 meter = 1000 millimeters (mm)
1 meter = 100 centimeters (cm)
1000 meters = 1 kilometer (km)

Volume   The amount of space an object takes up

liter (L)   A liter is slightly more than a quart.
1 liter = 1000 milliliters (mL)

Mass   The amount of matter in an object

gram (g)   A gram has a mass equal to about one paper clip.

1000 grams = 1 kilogram (kg)

Temperature   The measure of hotness or coldness

degrees   0°C = freezing point of water
Celsius (°C)   100°C = boiling point of water

Metric–English Equivalents

2.54 centimeters (cm) = 1 inch (in.)
1 meter (m) = 39.37 inches (in.)
1 kilometer (km) = 0.62 miles (mi)
1 liter (L) = 1.06 quarts (qt)
250 milliliters (mL) = 1 cup (c)
1 kilogram (kg) = 2.2 pounds (lb)
28.3 grams (g) = 1 ounce (oz)
°C = 5/9 x (°F – 32)

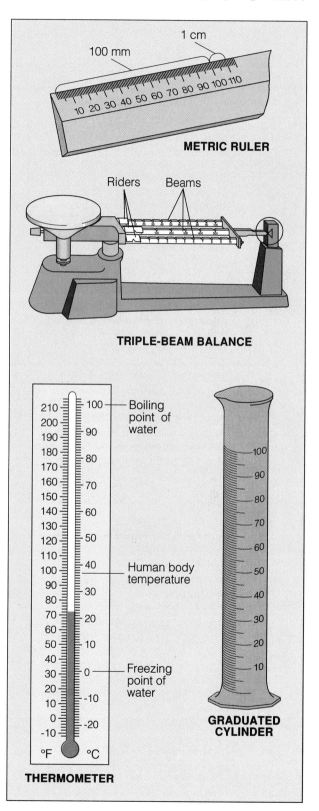

METRIC RULER

TRIPLE-BEAM BALANCE

THERMOMETER

GRADUATED CYLINDER

Glassware Safety

1. Whenever you see this symbol, you will know that you are working with glassware that can easily be broken. Take particular care to handle such glassware safely. And never use broken or chipped glassware.
2. Never heat glassware that is not thoroughly dry. Never pick up any glassware unless you are sure it is not hot. If it is hot, use heat-resistant gloves.
3. Always clean glassware thoroughly before putting it away.

Fire Safety

1. Whenever you see this symbol, you will know that you are working with fire. Never use any source of fire without wearing safety goggles.
2. Never heat anything—particularly chemicals—unless instructed to do so.
3. Never heat anything in a closed container.
4. Never reach across a flame.
5. Always use a clamp, tongs, or heat-resistant gloves to handle hot objects.
6. Always maintain a clean work area, particularly when using a flame.

Heat Safety

Whenever you see this symbol, you will know that you should put on heat-resistant gloves to avoid burning your hands.

Chemical Safety

1. Whenever you see this symbol, you will know that you are working with chemicals that could be hazardous.
2. Never smell any chemical directly from its container. Always use your hand to waft some of the odors from the top of the container toward your nose—and only when instructed to do so.
3. Never mix chemicals unless instructed to do so.
4. Never touch or taste any chemical unless instructed to do so.
5. Keep all lids closed when chemicals are not in use. Dispose of all chemicals as instructed by your teacher.

6. Immediately rinse with water any chemicals, particularly acids, that get on your skin and clothes. Then notify your teacher.

Eye and Face Safety

1. Whenever you see this symbol, you will know that you are performing an experiment in which you must take precautions to protect your eyes and face by wearing safety goggles.
2. When you are heating a test tube or bottle, always point it away from you and others. Chemicals can splash or boil out of a heated test tube.

Sharp Instrument Safety

1. Whenever you see this symbol, you will know that you are working with a sharp instrument.
2. Always use single-edged razors; double-edged razors are too dangerous.
3. Handle any sharp instrument with extreme care. Never cut any material toward you; always cut away from you.
4. Immediately notify your teacher if your skin is cut.

Electrical Safety

1. Whenever you see this symbol, you will know that you are using electricity in the laboratory.
2. Never use long extension cords to plug in any electrical device. Do not plug too many appliances into one socket or you may overload the socket and cause a fire.
3. Never touch an electrical appliance or outlet with wet hands.

Animal Safety

1. Whenever you see this symbol, you will know that you are working with live animals.
2. Do not cause pain, discomfort, or injury to an animal.
3. Follow your teacher's directions when handling animals. Wash your hands thoroughly after handling animals or their cages.

One of the first things a scientist learns is that working in the laboratory can be an exciting experience. But the laboratory can also be quite dangerous if proper safety rules are not followed at all times. To prepare yourself for a safe year in the laboratory, read over the following safety rules. Then read them a second time. Make sure you understand each rule. If you do not, ask your teacher to explain any rules you are unsure of.

Dress Code

1. Many materials in the laboratory can cause eye injury. To protect yourself from possible injury, wear safety goggles whenever you are working with chemicals, burners, or any substance that might get into your eyes. Never wear contact lenses in the laboratory.

2. Wear a laboratory apron or coat whenever you are working with chemicals or heated substances.

3. Tie back long hair to keep it away from any chemicals, burners and candles, or other laboratory equipment.

4. Remove or tie back any article of clothing or jewelry that can hang down and touch chemicals and flames.

General Safety Rules

5. Read all directions for an experiment several times. Follow the directions exactly as they are written. If you are in doubt about any part of the experiment, ask your teacher for assistance.

6. Never perform activities that are not authorized by your teacher. Obtain permission before "experimenting" on your own.

7. Never handle any equipment unless you have specific permission.

8. Take extreme care not to spill any material in the laboratory. If a spill occurs, immediately ask your teacher about the proper cleanup procedure. Never simply pour chemicals or other substances into the sink or trash container.

9. Never eat in the laboratory.

10. Wash your hands before and after each experiment.

First Aid

11. Immediately report all accidents, no matter how minor, to your teacher.

12. Learn what to do in case of specific accidents, such as getting acid in your eyes or on your skin. (Rinse acids from your body with lots of water.)

13. Become aware of the location of the first-aid kit. But your teacher should administer any required first aid due to injury. Or your teacher may send you to the school nurse or call a physician.

14. Know where and how to report an accident or fire. Find out the location of the fire extinguisher, phone, and fire alarm. Keep a list of important phone numbers—such as the fire department and the school nurse—near the phone. Immediately report any fires to your teacher.

Heating and Fire Safety

15. Again, never use a heat source, such as a candle or burner, without wearing safety goggles.

16. Never heat a chemical you are not instructed to heat. A chemical that is harmless when cool may be dangerous when heated.

17. Maintain a clean work area and keep all materials away from flames.

18. Never reach across a flame.

19. Make sure you know how to light a Bunsen burner. (Your teacher will demonstrate the proper procedure for lighting a burner.) If the flame leaps out of a burner toward you, immediately turn off the gas. Do not touch the burner. It may be hot. And never leave a lighted burner unattended!

20. When heating a test tube or bottle, always point it away from you and others. Chemicals can splash or boil out of a heated test tube.

21. Never heat a liquid in a closed container. The expanding gases produced may blow the container apart, injuring you or others.

22. Before picking up a container that has been heated, first hold the back of your hand near it. If you can feel the heat on the back of your hand, the container may be too hot to handle. Use a clamp or tongs when handling hot containers.

Using Chemicals Safely

23. Never mix chemicals for the "fun of it." You might produce a dangerous, possibly explosive substance.

24. Never touch, taste, or smell a chemical unless you are instructed by your teacher to do so. Many chemicals are poisonous. If you are instructed to note the fumes in an experiment, gently wave your hand over the opening of a container and direct the fumes toward your nose. Do not inhale the fumes directly from the container.

25. Use only those chemicals needed in the activity. Keep all lids closed when a chemical is not being used. Notify your teacher whenever chemicals are spilled.

26. Dispose of all chemicals as instructed by your teacher. To avoid contamination, never return chemicals to their original containers.

27. Be extra careful when working with acids or bases. Pour such chemicals over the sink, not over your workbench.

28. When diluting an acid, pour the acid into water. Never pour water into an acid.

29. Immediately rinse with water any acids that get on your skin or clothing. Then notify your teacher of any acid spill.

Using Glassware Safely

30. Never force glass tubing into a rubber stopper. A turning motion and lubricant will be helpful when inserting glass tubing into rubber stoppers or rubber tubing. Your teacher will demonstrate the proper way to insert glass tubing.

31. Never heat glassware that is not thoroughly dry. Use a wire screen to protect glassware from any flame.

32. Keep in mind that hot glassware will not appear hot. Never pick up glassware without first checking to see if it is hot. See #22.

33. If you are instructed to cut glass tubing, fire-polish the ends immediately to remove sharp edges.

34. Never use broken or chipped glassware. If glassware breaks, notify your teacher and dispose of the glassware in the proper trash container.

35. Never eat or drink from laboratory glassware. Thoroughly clean glassware before putting it away.

Using Sharp Instruments

36. Handle scalpels or razor blades with extreme care. Never cut material toward you; cut away from you.

37. Immediately notify your teacher if you cut your skin when working in the laboratory.

Animal Safety

38. No experiments that will cause pain, discomfort, or harm to mammals, birds, reptiles, fishes, and amphibians should be done in the classroom or at home.

39. Animals should be handled only if necessary. If an animal is excited or frightened, pregnant, feeding, or with its young, special handling is required.

40. Your teacher will instruct you as to how to handle each animal species that may be brought into the classroom.

41. Clean your hands thoroughly after handling animals or the cage containing animals.

End-of-Experiment Rules

42. After an experiment has been completed, clean up your work area and return all equipment to its proper place.

43. Wash your hands after every experiment.

44. Turn off all burners before leaving the laboratory. Check that the gas line leading to the burner is off as well.

Glossary

alga (AL-gah; plural: algae, AL-jee): any simple plantlike nonvascular photosynthetic autotroph

ameba (uh-MEE-bah): a bloblike protist that uses its pseudopods to move and to obtain food

angiosperm (AN-jee-oh-sperm): a plant whose seeds are contained in an ovary

annual: a plant that completes its life cycle within one growing season

antibiotic: a chemical that destroys or weakens harmful microorganisms

autotroph: an organism that obtains energy by making its own food

biennial: a plant that completes its life cycle in two years

binomial nomenclature (bigh-NOH-mee-uhl NOH-muhn-klay-cher): the naming system devised by Linnaeus in which each organism is given two names: a genus name and a species name

brown alga: an alga that contains brown acessory pigments and belongs to the phylum Phaeophyta

cilia (SIHL-ee-uh; singular: cilium): hairlike projections on the outside of cells that move with a wavelike beat; in ciliates, cilia help these organisms to move, obtain food, and sense the enviroment

ciliate (SIHL-ee-iht): an animallike protist that possesses cilia at some point in its life

cone: in plants, a reproductive structure of a gymnosperm

decomposer: an organism that breaks down dead organisms into simpler substances, thereby returning important materials to the soil and water

diatom (DIGH-ah-tahm): a plantlike protist that has a beautiful two-part glassy shell

dinoflagellate (digh-noh-FLAJ-eh-layt): a plantlike protist that typically has cell walls that look like plates of armor and possesses two flagella, one of which trails from one end like a tail and the other of which wraps around the middle of the organism like a belt

euglena (yoo-GLEE-nah): a plantlike flagellate protist that belongs to the genus *Euglena* and is characterized by a pouch that holds two flagella, a reddish eyespot, and a number of grass-green chloroplasts that are used in photosynthesis

flagellum (flah-JEHL-uhm; plural: flagella): a long whiplike structure that propels a cell through its environment

flower: a reproductive structure of an angiosperm

fruit: a ripened ovary of a plant that encloses and protects the seed or seeds

fungus (FUHN-guhs; plural: fungi, FUHN-jigh): a heterotroph, usually multicellular, that releases chemicals that digest the substance on which it is growing and then absorbs the digested food; multicellular fungi are made up of threadlike hyphae; many fungi reproduce by means of spores

genus (plural: genera): the second-smallest taxonomic group; a genus consists of a number of similar, closely related species; a number of closely related genera make up a family

green alga: an alga that gets its color from the green pigment chlorophyll and belongs to the phylum Chlorophyta

gymnosperm (JIHM-noh-sperm): a plant whose seeds are not contained in an ovary

heterotroph: an organism that cannot make its own food and thus must eat other organisms in order to obtain energy

host: a living thing that provides a home and/or food for a parasite

hypha (plural: hyphae, HIGH-fee): one of the branching, threadlike tubes that makes up the body of a multicellular fungus

leaf: a structure in vascular plants whose main function is typically photosynthesis

lichen (LIGH-kuhn): a plantlike structure that is formed by a fungus and an alga that live together

mold: a fuzzy, shapeless, fairly flat multicellular fungus

mushroom: a multicellular fungus shaped like an umbrella

nonvascular plant: a plant that lacks vascular tissue

ovule (AH-vyool): a structure in a female cone or flower that contains an eggs cell and develops into a seed

paramecium (par-uh-MEE-see-uhm; plural: paramecia): a slipper-shaped ciliate protist that belongs to the genus *Paramecium*

parasite (PAIR-ah-sight): an organism that survives by living on or in a host organism, thus harming it

perennial: a plant that lives for many years

petal: a colorful leaflike flower structure that serves to attract pollinators

phloem (FLOH-ehm): plant vascular tissue that carries food

photosynthesis (foht-oh-SIHN-thuh-sihs): the food-making process that involves chlorophyll, in which light energy is used to make food (glucose) from carbon dioxide and water

pigment: a colored chemical; plant pigments are often associated with photosynthesis

pistil (PIHS-tihl): a female reproductive organ in a flower, which consists of a stigma, style, and ovule-containing ovary

pollen (PAHL-uhn): tiny grains that can be thought of as containing sperm cells, which are produced by male cones and flowers

pollination (pahl-uh-NAY-shuhn): the process by which pollen is carried from male reproductive structures to female reproductive structures

protist: a unicellular organism that contains a nucleus

pseudopod (SOO-doh-pahd): a temporary extension of the cell membrane and cytoplasm used in feeding and/or movement

red alga: an alga that contains red accessory pigments and belongs to the phylum Rhodophyta

root: a structure in vascular plants whose main functions are typically absorption and anchorage

sarcodine (SAHR-koh-dighn): an animallike protist that possesses pseudopods

sepal (SEE-puhl): a leaflike structure that protects a developing flower

slime mold: a funguslike protist that is a microscopic amebalike cell at one stage of its life cycle and a large, moist, flat, shapeless blob at another stage; slime molds produce reproductive structures known as fruiting bodies, which contain spores

species: the smallest and most specific taxonomic group, which consists of individuals that are quite similar in appearance and behavior and that can interbreed to produce fertile offspring

spore: a cell, usually surrounded by a protective wall, that is specialized either for reproduction or for a resting stage; the reproductive cells produced by most fungi are known as spores

sporozoan (spohr-oh-ZOH-uhn): a parasitic animallike protist that has a complex life cycle involving more than one kind of host animal and that typically produces cells called spores in order to pass from one host to another

stamen (STAY-muhn): a male reproductive organ in a flower, which consists of a filament topped by a pollen-producing anther

stem: a structure in vascular plants whose main functions are typically to carry materials between the roots and leaves and to support the plant

symbiosis (sihm-bigh-OH-sihs; plural: symbioses): a relationship in which one organism lives on, near, or even inside another organism and at least one of the organisms benefits

tropism (TROH-pihz-uhm): in plants, the growth of a plant toward or away from a stimulus

vascular plant: a plant that has vascular tissue

virus: a disease-causing particle consisting of hereditary material enclosed in a protein coat that is smaller and less complex than a cell

xylem (ZIGH-luhm): plant vascular tissue that carries water and minerals from the roots up through a plant and that also helps to support a plant

yeast: a unicellular fungus

zooflagellate (zoh-oh-FLAJ-ehl-iht): an animallike protist that possesses flagella

Index

Credits

Cover Background: Ken Karp
Photo Research: Natalie Goldstein
Contributing Artists: Michael Adams/ Phil Veloric, Art Representatives; Ray Smith; Warren Budd Associates Ltd.; Fran Milner; Function Thru Form; Keith Kasnot; David Biedrzycki; Gerry Schrenk; Carol Schwartz/Dilys Evans, Art Representatives
Photographs: 5 left: Dwight Kuhn Photography; right: D. Avon/Ardea London; **6** top: Lefever/Grushow/Grant Heilman Photography; center: Index Stock Photography, Inc.; bottom: Rex Joseph; **8** left: London School of Hygiene & Tropical Medicine/SPL/Photo Researchers, Inc.; right: Dr. Ann Smith/SPL/Photo Researchers, Inc.; **9** left: S. Nielsen/DRK Photo; right: D. Cavagnaro/DRK Photo; **10** and **11** Peter Scoones/Seaphoto Ltd./Planet Earth Pictures; **14** top and center: Breck P. Kent; bottom: Robert Frerck/Odyssey Productions; **15** left: Lawrence Migdale/Photo Researchers, Inc.; right: Dan Guravich/Photo Researchers, Inc.; **16** left: Stephen Dalton/ Animals Animals/Earth Scenes; right: M. P. Kahl/DRK Photo; **17** Breck P. Kent; **18** top: E.R. Degginger/Animals Animals/Earth Scenes; bottom: Stephen J. Krasemann/ DRK Photo; **20** top left: Barbara J. Wright/ Animals Animals/Earth Scenes; top center: Robert C. Simpson/Tom Stack & Associates; top right: Charles Palek/Tom Stack & Associates; bottom: Tom & Pat Leeson/DRK Photo; **21** D. Avon/Ardea London; **22** left to right, top to bottom: Phil Dotson/DPI; Jack Dermid; Larry Lipsky/ DRK Photo; T. Zywotko/DPI; Larry Roberts/ Visuals Unlimited; David M. Stone; Barbara K. Deans/DPI; T. Zywotko/DPI; Jerry Frank/DPI; DPI; J. Alex Langley/DPI; T. Zywotko/DPI; Phil Dotson/DPI; Phil Dotson/DPI; F. Erizel/Bruce Coleman, Inc.; T. Zywotko/DPI; Lois and George Cox/Bruce Coleman, Inc.; Marty Stouffer/Animals Animals/Earth Scenes; Mimi Forsyth/Monkmeyer Press; T. Zywotko/DPI; Kenneth W. Fink/Bruce Coleman, Inc.; Wil Blanche/ DPI; Phil Dotson/DPI; T. Zywotko/DPI; Jon A. Hull/Bruce Coleman, Inc.; **26** top and bottom: Peter Parks/Oxford Scientific Films/Animals Animals/Earth Scenes; center: T. E. Adams/Visuals Unlimited; **27** left: Stephen J. Krasemann/DRK Photo; top right: M & A Doolittle/Rainbow; bottom right: Michael Fogden/DRK Photo; **28** top: Michael Fogden/DRK Photo; bottom: G.I. Bernard/Oxford Scientific Films/Animals Animals/Earth Scenes; **29** top: Michael Dick/ Animals Animals/Earth Scenes; bottom: Richard Shiell/Animals Animals/Earth Scenes; **33** Michael Fogden/DRK Photo; **34** and **35** Biozentrum/Science Photo Library/Photo Researchers, Inc.; **36** top left: CNRI/Science Photo Library/Photo Researchers, Inc.; bottom left: Tektoff-RM/ CNRI/Science Photo Library/Photo Researchers, Inc.; bottom center: A. B. Dowsett/Science Photo Library/Photo Researchers, Inc.; bottom right: Dr. Oscar Bradfute/Peter Arnold, Inc.; **37** A. Jones/Visuals Unlimited; **39** Lee D. Simon/Photo Researchers, Inc.; **40** top: CNRI/Science Photo Library/Photo Researchers, Inc.; bottom: Tom Broker/Rainbow; **41** top left: Ed Reschke/Peter Arnold, Inc.; top right: Tektoff-RM/CNRI/Science Photo Library/ Photo Researchers, Inc.; bottom: CNRI/Science Photo Library/Photo Researchers, Inc.; **42** top: Sepp Seitz/Woodfin Camp & Associates; bottom left: Renee Lynn/Photo Researchers, Inc.; bottom right: Chuck O'Rear/Woodfin Camp & Associates; **43** top and bottom right: CNRI/Science Photo Library/Photo Researchers, Inc.; left: David M. Phillips/Visuals Unlimited; center: A. B. Dowsett/Science Photo Library/Photo Researchers, Inc.; **44** left: David M. Phillips/Visuals Unlimited; right: Biophoto Associates/Photo Researchers, Inc.; **45** top: Dr. Tony Brain/Science Photo Library/Photo Researchers, Inc.; bottom: A. B. Dowsett/Science Photo Library/Photo Researchers, Inc.; **46** top: Wolfgang Kaehler; bottom left: Alfred Pasieka/Science Photo Library/Photo Researchers, Inc.; bottom right: Dr. L. Caro/ Science Photo Library/Photo Researchers, Inc.; **47** left: John Cancalosi/DRK Photo; right: Alfred Pasieka/Science Photo Library/Photo Researchers, Inc.; **48** Tim Rock/Animals Animals/Earth Scenes; **49** top: Hugh Spencer/Photo Researchers, Inc.; bottom: Dr. Jeremy Burgess/Science Photo Library/Photo Researchers, Inc.; **50** top: Don & Pat Valenti/DRK Photo; bottom: B. Nation/Sygma; **51** Bonnie Sue Rauch/ Photo Researchers, Inc.; **52** top: Martin M. Rotker; bottom: Cabisco/Visuals Unlimited; **53** Dr. R. Clinton Fuller/University of Massachusetts; **58** and **59** London School of Hygiene & Tropical Medicine/Science Photo Library/Photo Researchers, Inc.; **60** left: Cabisco/Visuals Unlimited; center: A. M. Siegelman/Visuals Unlimited; right: K. G. Murti/Visuals Unlimited; **61** top: T. E. Adams/Visuals Unlimited; bottom: Cabisco/Visuals Unlimited; **62** left: A. M. Siegelman/Visuals Unlimited; right: Biophoto Associates/Photo Researchers, Inc.; **63** top: R. Oldfield/Polaroid/Visuals Unlimited; bottom: A. M. Siegelman/Visuals Unlimited; **64** left: David M. Phillips/Visuals Unlimited; right: Omikron/Science Source/Photo Researchers, Inc.; **65** top: Cabisco/Visuals Unlimited; bottom left and right: Michael Abbey/Visuals Unlimited; **66** Karl Aufderheide/Visuals Unlimited; **67** top: Michael Abbey/Photo Researchers, Inc.; bottom left: Bruce Iverson/Visuals Unlimited; bottom right: CBS/Visuals Unlimited; **68** top: David M. Phillips/Visuals Unlimited; bottom: Jerome Paulin/Visuals Unlimited; **69** top left: Michael Abbey/Visuals Unlimited; top right: Raymond A. Mendez/Animals Animals/Earth Scenes; bottom: Will & Deni McIntyre/Photo Researchers, Inc.; **71** left: © Lennart Nilsson, THE INCREDIBLE MACHINE, National Geographic Society; right: © Lennart Nilsson, National Geographic Society; **72** top: G. I. Bernard/Animals Animals/Earth Scenes; bottom: SCIENCE, "Predator-Induced Trophic Shift of a Free-Living Ciliate: Parasitism of Mosquito Larvae by Their Prey" published 05/27/88. Volume 240, beginning on page 1193. Dr. Jan O. Washburn, University of California, Berkeley. © 1988 by the AAAS.; **73** top right and bottom left and right: Dr. John C. Steinmetz; **74** top: T. E. Adams/Visuals Unlimited; bottom: David M. Phillips/ Visuals Unlimited; **75** top left and right: Cabisco/Visuals Unlimited; top center: Cecil Fox/Science Source/Photo Researchers, Inc.; bottom left: Veronika Burmeister/Visuals Unlimited; right: Nuridsany et Pérennou/Photo Researchers, Inc.; **76** top: Dr. J.A.L. Cooke/Oxford Scientific Films/Animals Animals/Earth Scenes; bottom: David M. Phillips/Visuals Unlimited; **77** left: Biophoto Associates/Science Source/Photo Researchers, Inc.; top right: Peter Parks/Oxford Scientific Films/Animals Animals/Earth Scenes; bottom right: M.I. Walker/Science Source/Photo Researchers, Inc.; **78** top: Dwight Kuhn Photography; bottom left: Cabisco/Visuals Unlimited; bottom right: CBS/Visuals Unlimited; **79** top left: V. Duran/Visuals Unlimited; top right: G. I. Bernard/Oxford Scientific Films/Animals Animals/Earth Scenes; center: Cabisco/Visuals Unlimited; bottom: CBS/Visuals Unlimited; **83** David M. Phillips/Visuals Unlimited; **84** and **85** Adam Woolfitt/Woodfin Camp & Associates; **86** top and bottom left: Michael Fogden/DRK Photo; bottom right: Charles Brewer; **87** top left: John Gerlach/DRK Photo; top right: S. Flegler/Visuals Unlimited; center right: Michael Fogden/Oxford Scientific Films/Animals Animals/Earth Scenes; bottom right: Biophoto Associates/Science Source/Photo Researchers, Inc.; **88** left: Don & Pat Valenti/DRK Photo; center: S. Nielsen/DRK Photo; right: C. Gerald Van Dyke/Visuals Unlimited; **89** top left: Michael Fogden /Oxford Scientific Films/Animals Animals/Earth Scenes; center: Michael Fogden/DRK Photo; right: P. L. Kaltreider/Visuals Unlimited; bottom left: Cabisco/Visuals Unlimited; **90** John Gerlach/DRK Photo; **91** top: Michael Fogden/Oxford Scientific Films/Animals Animals/Earth Scenes; bottom: Dick Poe/Visuals Unlimited; **92** J. Forsdyke/ Gene Cox/Science Photo Library/Photo Researchers, Inc.; **93** left: Andrew McClenaghan/Science Photo Library/Photo Researchers, Inc.; right: Dr. Jeremy Burgess/Science Photo Library/Photo Researchers, Inc.; **95** top left: William D. Griffin/Animals Animals/Earth Scenes; top right: Richard K. LaVal/Animals Animals/Earth Scenes; bottom: D. Cavagnaro/DRK Photo; **96** top: Patti Murray/ Animals Animals/Earth Scenes; bottom: Glenn Oliver/Visuals Unlimited; **97** left: Tom Bean/DRK Photo; right: Breck P. Kent; **98** Cary Wolinsky/Stock Boston, Inc.; **99** top left: Bates Littlehales/Animals Animals/Earth Scenes; top right: Stephen J. Krasemann/DRK Photo; bottom left: E. R. Degginger/Animals Animals/Earth Scenes; bottom right: Larry Ulrich/DRK Photo; **103** Charlie Palek/Animals Animals/Earth Scenes; **104** and **105** Jeffrey L. Rotman; **106** top: L. L. Sims/Visuals Unlimited; bottom left: Anne Wertheim/Animals Animals/Earth Scenes; bottom right: Robert & Linda Mitchell Photography; **107** top right: Biophoto Associates/Science Source/Photo Researchers, Inc.; center right: M. I. Walker/Science Source/Photo Researchers, Inc.; bottom left: Jeffrey L. Rotman; bottom center: Robert & Linda Mitchell Photography; bottom right: Charles Seaborn/Odyssey Productions; **109** top: Runk/Schoenberger/Grant Heilman Photography; bottom left: Heather Angel; bottom right: Robert Maier/Animals Animals/ Earth Scenes; **110** top left: Robert & Linda Mitchell Photography; top right: Charles Seaborn/Odyssey Productions; bottom: Doug Wechsler/Animals Animals/Earth Scenes; **111** top left: Jeffrey L. Rotman; top right: Jack Dermid; center right: Charles Seaborn/Odyssey Productions; bottom right: John Durham/Science Photo Library/Photo Researchers, Inc.; **112** top: Robert Frerck/Odyssey Productions; center: Charles Seaborn/Odyssey Productions; bottom: Dwight Kuhn Photography; **113** top left: M. I. Walker/Photo Researchers, Inc.; top right: Jeff Foott Productions; bottom left: Oxford Scientific Films/Animals Animals/Earth Scenes; bottom right: Charles Seaborn/Odyssey Productions; **116** top: Doug Wechsler/Animals Animals/Earth Scenes; bottom left: Robert & Linda Mitchell Photography; bottom right: Frans Lanting/Minden Pictures, Inc.; **117** Dwight Kuhn Photography; **118** top: David Muench Photography Inc.; bottom: Stephen J. Krasemann/DRK Photo; **119** left: Kjell B. Sandved; right: Jack Dermid; **120** top left: Stephen J. Krasemann/DRK Photo; top right and bottom left: Kjell B. Sandved; bottom right: Wolfgang Kaehler; **121** Robert & Linda Mitchell Photography; **122** top left: Doug Wechsler/Animals Animals/Earth Scenes; top right, bottom left and right: Kjell B. Sandved; **123** Garry Gay/Image Bank; **127** Robert & Linda Mitchell Photography; **128** and **129** John Lemker/Animals Animals/Earth Scenes; **130** top: Wolfgang Kaehler; bottom left: Robert & Linda Mitchell Photography; bottom right: Michael Fogden/DRK Photo; **132** top: P. Dayanandan/Photo Researchers, Inc.; center: J. F. Gennaro/Photo Researchers, Inc.; bottom: Robert & Linda Mitchell Photography; **133** top left: Jack Swenson/Tom Stack & Associates; top center: Wolfgang Kaehler; top right: Robert & Linda Mitchell Photography; bottom left: Kjell B. Sandved; bottom center and bottom right: Dwight Kuhn Photography; **134** top: Frans Lanting/Minden Pictures, Inc.; center: Jim Brandenburg/Minden Pictures, Inc.; bottom: T. A. Wiewandt/DRK Photo; **135** top: Robert & Linda Mitchell Photography; bottom: Ardea London; **136** top left and right: Robert & Linda Mitchell Photography; bottom left: Jeff Foott/Tom Stack & Associates; bottom center: Dwight Kuhn Photography; bottom right: D. Cavagnaro/DRK Photo; **137** left and center left: Kjell B. Sandved; center right: Wolfgang Kaehler; top right and bottom right: Robert & Linda Mitchell Photography; **138** DPI; **140** left and right: Dr. Jeremy Burgess/Science Photo Library/Photo Researchers, Inc.; **141** top left: Breck P. Kent; top center and top right: Robert & Linda Mitchell Photography; bottom left: Jeff Foott/Tom Stack & Associates; bottom center: Dwight Kuhn Photography; bottom right: Kjell B. Sandved; **142** left: Tom Bean/DRK Photo; right: Dwight R. Kuhn/DRK Photo; **143** left: Robert & Linda Mitchell Photography; top right and bottom right: Wolfgang Kaehler; **144** top left: Dr. Jeremy Burgess/ Science Photo Library/Photo Researchers, Inc.; right top, center bottom, and bottom right: Robert& Linda Mitchell Photography; center top: Patti Murray/Animals Animals/Earth Scenes; bottom left: CNRI/ Science Photo Library/Photo Researchers, Inc.; **145** left, center, and right: Kjell B. Sandved; **146** left and bottom right: Robert & Linda Mitchell Photography; top right: Jeff Foott/Tom Stack & Associates; center: Adrienne T. Gibson/Animals Animals/ Earth Scenes; **147** top left: Mickey Gibson/Animals Animals/Earth Scenes; top right: Robert & Linda Mitchell Photography; bottom: Coco McCoy/Rainbow; **148** top: Kjell B. Sandved; bottom: E. R. Degginger/Animals Animals/Earth Scenes; **149** top left: William E. Ferguson; top center and top right: Wolfgang Kaehler; bottom: Peter Pickford/DRK Photo; **150** top left: D. Cavagnaro/DRK Photo; top right and bottom left: David Muench Photography Inc.; **151** top left: Michael Fogden/ DRK Photo; top right: J.A.L. Cooke/Oxford Scientific Films/Animals Animals/Earth Scenes; center right: G. I. Bernard/Oxford Scientific Films/Animals Animals/Earth Scenes; bottom left: Alastair Shay/Oxford Scientific Films/Animals Animals/Earth Scenes; bottom right: Stanley Breeden/ DRK Photo; **154** top left: William E. Ferguson; right: John Gerlach/DRK Photo; bottom left: Don & Pat Valenti/DRK Photo; **155** Runk/Schoenberger/Grant Heilman Photography; **159** Michael J. Doolittle/ Rainbow; **160** Michael Elias; **161** top and center: Dudley Foster/Woods Hole Oceanographic Institution; bottom: Colleen Cavanaugh; **162** David Scharf/Peter Arnold, Inc.; **163** Annie Griffiths/DRK Photo; **164** Michael Habicht/Animals Animals/Earth Scenes; **166** Terrence Moore/Woodfin Camp & Associates; **167** Runk/Schoenberger/Grant Heilman Photography; **168** Tom & Pat Leeson/DRK Photo; **189** Kjell B. Sandved; **191** D. Cavagnaro/DRK Photo